Praise for *Improbable Destinies*

"With an ideal combination of clarity and comedy, scholarly caution and infectious enthusiasm, Losos shows us how evolutionary biology opens up for each of us the glorious workings of our world, with surprises around every corner."

—*The Washington Post*

"This is a wonderfully serious book with a lighthearted voice. Is evolution predictable or contingent? Big question. Why do adaptations converge? Big question. Why is the platypus unique? Smaller question, but fun! Read, enjoy, think."

—David Quammen, author of *The Song of the Dodo* and *Spillover*

"Packed with stories of capturing lizards in the field, *Improbable Destinies* explores how we think evolutionary changes happen in populations, from mice to microbes to sticklebacks. Get this for the backyard biologist in your life."

—*Popular Science*

"Deep, broad, brilliant and thought-provoking . . . In staggeringly clear and engaging prose, Losos shows us remarkable vignettes of scientists working at personal and professional risk in all sorts of habitats—field, lab and museum—to elucidate stunning mechanisms of evolution. . . . He is one of the premier writers in biology today."

—*Nature*

"*Improbable Destinies* is one of the best books on evolutionary biology for a broad readership ever written. Its subjects—the unfolding of Earth's biological history, the precarious nature of human existence, and the likelihood of life on exoplanets—are presented in a detailed, exciting style expected from an authentic scientist and naturalist."

—Edward O. Wilson, University Research
Professor Emeritus, Harvard University

"Losos deals with some of the great questions about evolution. . . . As he muses on these issues, [Losos] recounts field stories, complete with hilarious misadventures. He introduces an esoteric cast of scientific characters obsessed with moths, fish, lizards, invertebrates and other creatures, each revealing something new and interesting about evolution itself." —*The Wall Street Journal*

"[A] compelling book." —*Science*

"Is evolution a story foretold? Or is it little more than the rolls of DNA's dice? In *Improbable Destinies*, Jonathan Losos tackles these fascinating questions not with empty philosophizing, but with juicy tales from the front lines of scientific research. Drunk flies, fast-evolving lizards, mutating microbes, and hypothetical humanoid dinosaurs all grace the pages of this wonderfully thought-provoking book." —Carl Zimmer, author of *A Planet of Viruses* and *The Tangled Bank*

"A rich, provocative, and very accessible book, *Improbable Destinies* is an exclusive behind-the-scenes tour of the ecological theater and evolutionary play of life, expertly guided by one of its most insightful observers. Jonathan Losos has shone a light on a largely unheralded cast of fascinating creatures and ingenious scientists who are reshaping our view of why life is the way it is."
—Sean B. Carroll, author of *The Serengeti Rules* and *Brave Genius*

"Losos writes accessibly and interestingly about recent research in evolutionary biology. Today's scientists have figured out ways to prove that evolution doesn't demand the millennia Darwin assumed."
—*St. Louis Post-Dispatch*

"In a refreshingly accessible narrative, laced with piquant anecdotes, Losos underscores the human significance of science affecting not only how we interpret our own place on the planet but also how we envision life in distant galaxies. Wonderfully lucid; singularly engaging."
—*Booklist* (starred review)

"A thoroughly accessible analysis of whether evolution is one big crapshoot or rather mundanely predictable. No spoilers here, but the evidence presented on both sides makes for some thought-provoking reading."
—*Washington Independent Review of Books*

"A cheerful, delightfully lucid primer on evolution and the predictive possibilities within the field."
—*Kirkus Reviews* (starred review)

"Every now and then a brilliant book comes along that helps us rethink what we know about a subject. Jonathan B. Losos's fascinating, compulsively readable *Improbable Destinies* is just such a book. . . . With vivacious writing and thoughtful, provocative insights, Losos's captivating study of evolution deserves to be read alongside the books of E. O. Wilson (*The Social Conquest of Earth*) and Stephen Jay Gould (*Wonderful Life*)."
—*BookPage*

"Losos explains both the science and the underlying philosophy of the questions being asked in an accessible and engaging manner. . . . The book is as enjoyable as it is informative."
—*Publishers Weekly*

"*Improbable Destinies* is a crackling good read, threading rich anecdote into trenchant science. It belongs on the same shelf as *I Contain Multitudes*, Ed Yong's gorgeously crafted account of microbes and their critical roles in our bodies; Nick Lane's dense, groundbreaking work on the origins of life, *The Vital Question*; and other recent books that grapple with Darwin's revolution, such as Richard O. Prum's *The Evolution of Beauty* and Robert M. Sapolsky's *Behave*."
—*The Barnes & Noble Review*

IMPROBABLE DESTINIES

Fate, Chance, and the Future of Evolution

JONATHAN B. LOSOS

Illustrated by Marlin Peterson

Riverhead Books
New York

RIVERHEAD BOOKS
An imprint of Penguin Random House LLC
375 Hudson Street
New York, New York 10014

Illustration credits:
Pages 7, 11, 15, 19, 35, 36, 42, 43, 47, 49, 59, 61, 67, 74, 84, 85,
90, 102, 104, 113, 119, 126, 128, 135, 156, 160, 170, 188, 192,
197, 203, 206, 221, 236, 243, 275, 316, 325, 329 copyright © by
Marlin Peterson; page 82 copyright © by David Tuss

The Library of Congress has catalogued the Riverhead
hardcover edition as follows:

Names: Losos, Jonathan B.
Title: Improbable destinies : fate, chance, and the future of
    evolution / Jonathan B. Losos.
Description: New York : Riverhead Books, 2017. | Includes
    bibliographical references and index.
Identifiers: LCCN 2016054594 | ISBN 9780399184925
Subjects: LCSH: Evolution (Biology)
Classification: LCC QH366.2 .L664 2017 | DDC 576.8—dc23
LC record available at https://lccn.loc.gov/2016054594

First Riverhead hardcover edition: August 2017
First Riverhead trade paperback edition: August 2018
Riverhead trade paperback ISBN: 9780525534136

Book design by Lauren Kolm

147468846

To my wife, Melissa Losos, and my parents, Joseph and Carolyn Losos, for their love and support

Contents

PART THREE · EVOLUTION UNDER THE MICROSCOPE

Preface

L ike many children, I went through a dinosaur phase. I was legendary at nursery school for showing up every day with my basket full of plastic dinosaurs: *Allosaurus, Stegosaurus, Ankylosaurus, Tyrannosaurus rex.* I had 'em all, or at least all of the twenty or so species available back then (kids today have it so much better).

Unlike most kids, I never grew out of the phase. I still have my toy dinos, many more now; I still know their names, can even still pronounce *Parasaurolophus* (pair-uh-soar-ahl-oh-fuss). But my interests have shifted to living reptiles: snakes, turtles, lizards, and crocodilians.

To a large extent, this shift was prompted by a rerun of the old TV show *Leave It to Beaver,* specifically the episode in which Wally and the Beave buy a mail-order baby alligator and hide it in the bathroom. Needless to say, when Minerva the housekeeper finds it, hilarity ensues. Thinking that this was a great idea and knowing that pet stores in those days (the early 1970s) sold baby caimans, the Central and South American version of the alligator, I petitioned my mother. Not being the sort of woman who likes to say no, she suggested we contact a family friend, Charlie Hoessle, the deputy director of the Saint Louis Zoo, expecting him to put the kibosh on the idea. Every day when my father returned home from work, my first question was "Did you talk to Mr. Hoessle today?" By nature not especially patient (particularly not at age ten), I went from exasperation to annoyance to anger as the days passed. What was the problem? My father was waiting to see Hoessle at a meeting, rather than simply call him up. Would that meeting never happen? Just as I was giving up all hope of a crocodilian-in-residence, my father came home one evening and announced that he'd spoken with Mr. Hoessle. "And the answer?" I asked, as I fidgeted with hope and nervousness. Then elation: Hoessle said that it was a great idea, the same way he got his start in herpetology!* My mother was stuck, and soon our basement was full of all manner of reptile. I was on my way to my own career in the field.

At the same time that I was tending my scaly charges, I was also an avid reader of the monthly magazine *Natural History*, put out by the American Museum of Natural History in New York. A highlight of every issue was the column This View of Life by the brilliant and erudite Harvard paleontologist Stephen Jay Gould. Its name lifted from the closing sentence of Darwin's *On the Origin of Species*, the column regularly explored Gould's heterodox ideas about the evolutionary process, often stressing the indeterminate and unpredictable nature of evolution. Elegantly written and mixing in vignettes from history, architecture, and baseball, Gould presented a compelling case for his worldview.

When I was accepted to Harvard in 1980, I looked forward to learning from the great man himself in his modestly entitled non-majors course, The History of Earth and of Life. And fascinating he was, as engaging in person as he was in print. But the faculty member who made the greatest impression on me was Ernest Williams, the

---

* The study of amphibians and reptiles.

Curator of Herpetology at Harvard's Museum of Comparative Zoology (the position I now hold). An imperious elder scientist, he nonetheless was very welcoming to a young upstart with an interest in reptiles. Soon I found myself studying the particular type of lizards that had been the focus of his life's work.

Small, usually green or brown, with sticky pads on their toes and an extensible flap of colorful skin under their throats, anole lizards are photogenic and captivating in their antics. But what has shot them to fame in scientific circles is their evolutionary exuberance. Four hundred species are known, and more are being discovered every year, making *Anolis* one of the largest genera of vertebrate animals. This immense diversity is accomplished by great local richness—as many as a dozen or more co-occurring species—combined with regional endemicity, most species being limited to a single island or small part of the tropical American mainland.

In the 1960s, Williams' graduate student Stan Rand documented that anole species coexist by adapting to different parts of the habitat, some living high in the tree, others in the grass or on twigs. Williams' great insight was to realize that the same set of habitat specialists had evolved on each island in the Greater Antilles (Cuba, Hispaniola, Jamaica, and Puerto Rico). That is, the lizards had diversified independently, yet had evolved to divvy up the available habitats in almost exactly the same way on all four islands.

As an undergraduate, I worked on a small part of this story, conducting an honor's research project on the interactions between two species in the Dominican Republic. I graduated and headed to a Ph.D. program in California, vowing never to work on these lizards again because everything important had already been discovered by Williams and his lab.

Ah, the naïveté of youth. As anyone who's done science knows, successful projects usually answer one question, but lead to three new ones. It took me two years of graduate school and a dozen failed projects, but I finally realized island anoles are a perfect group for studying how evolutionary diversification occurs.

So I spent four years traipsing through the Caribbean, climbing trees, catching lizards, and sipping the occasional piña colada. And by the end, I had shown, using the latest analytical techniques, that Williams was exactly right. Anatomically and ecologically very similar species had evolved independently on the different islands. Moreover, my studies of biomechanics—how the lizards run, jump, and cling—revealed the adaptive basis for anatomical variation, explaining why features such as long legs or big toepads evolved for species using particular parts of the habitat.

The ink was barely dry on my dissertation when *Wonderful Life: The Burgess Shale and the Nature of History*, arguably Stephen Jay Gould's greatest work, appeared in bookstores. I read it voraciously and found the argument persuasive. Evolution's path is quirky and unpredictable, he argued; replay life's tape again and you'd get a very different outcome.

But hold on. Gould's conceit of turning back the clock and replaying the evolutionary tape of life is impossible (at least in nature), but another way to test the repeatability of evolution would be to play the same tape in multiple locations. Aren't Caribbean islands, each seeded with an ancestral anole lizard, essentially the same as replaying the tape of life? Assuming that the islands have more or less the same environments, doesn't this constitute a test of evolutionary repeatability?

Indeed it does, and I found myself in an intellectual conundrum. Gould convincingly argued that evolution shouldn't repeat itself, yet my own research showed that it did. Was Gould wrong, or was my work the exception that somehow proved the rule? I opted for the latter explanation, embracing the Gouldian worldview even as my own work provided a counterexample.

The last quarter century has been challenging for this perspective. An intellectual counterpoint to Gould's emphasis on unpredictability and non-repeatability has emerged. This alternative view emphasizes the ubiquity of adaptive convergent evolution: species living in similar environments will evolve similar features as adaptations to the shared natural selection pressures they experience. My *Anolis* lizards are an example of such convergence. Proponents of this view argue that convergence demonstrates that evolution, far from being quirky and indeterminate, is actually quite predictable: there are limited ways to make a living in the natural world, so natural selection drives the evolution of the same features time and time again.

Evolutionary biology has advanced considerably since *Wonderful Life* was published and I received my doctorate. New ideas, new approaches, and new methods of data collection have emerged. The number of scientists studying evolution has increased enormously. We've cracked the genome, mapped the tree of life, learned about the evolving microbiome. Spectacular fossil discoveries have clarified much of evolution's history.

These data have much to say about the predictability of evolution. The more we learn about life's history on this planet, the more we see that convergence has occurred, that very similar outcomes have evolved repeatedly. My anoles seem less exceptional, Gould's rule more in doubt.

But we now know that there's another way to study evolution besides documenting what has happened through the ages. We've discovered that we can study evolution as it occurs, right before our eyes. And that means that we actually *can* replay the tape by harnessing the power of the experimental method—the hallmark of laboratory science—to address the question of evolutionary predictability.

Experiments are a powerful way to study evolution. And they're also a lot of fun. You may remember experiments from high school chemistry class. Mixing chemical reagents in beakers and pouring them into test tubes was not particularly enjoyable—at least it wasn't for me. But when your test tubes are Bahamian islands and your reagents are lizards, it's a completely different story. Sure, sometimes the Sun is a bit strong, and there's nothing more frustrating than failing to catch an important lizard because you were distracted by a dolphin swimming by. But experimental evolution is at the cutting edge of evolutionary biology, allowing us to actually test our ideas about evolution, out in nature, in real time. What could be more exciting? Evolution experiments are now going on all over the world—from the montane rainforests of Trinidad to the Sandhills of Nebraska to British Columbian ponds—and they're letting us directly investigate whether evolution can be predicted.

Oh to be a graduate student again. It's a glorious time to be an evolutionary biologist, a golden era. With the tools available, from genome sequencing to field experiments, we can finally answer the questions that have bedeviled our field for the last century.

I set out to write a book about the ongoing work to answer one such question: how predictable is evolution? But while writing, I discovered that this book had to be about much more than just what the science is telling us. Scientific knowledge doesn't just appear out of nowhere; it's a result of scientists toiling away, using their creativity and insight to learn about the natural world. And the people studying evolutionary predictability are a particularly fascinating lot.

In this light, *Improbable Destinies* will be not just what we know about evolution, but how we know what we know. Not just the technology and theories of science, but where the ideas come from—how researchers think them up, how they are honed by experiences in the field, how much of science is the serendipitous juxtaposition of disparate ideas brought together by unexpected observations. Moreover, the seemingly esoteric academic questions they study turn out to actually matter, both to our understanding of our own place in the universe and how life around us is coping with a changing world. As a result, *Improbable Destinies* is a story of people and places, plants and animals, big questions and pressing problems. And it begins, like my love of the natural world, with dinosaurs.

The Good Dinosaur

The trailer for the Pixar movie *The Good Dinosaur* begins with an asteroid belt packed full of oversized boulders. One asteroid shoots through the rock pile, slamming into another, which ricochets into a third, sending it zooming off into space, straight toward a distant object. As the object gets larger, its identity becomes obvious: a blue planet with patches of green and wisps of white. "Millions of years ago, an asteroid six miles wide destroyed every dinosaur on Earth," the narrator intones. We see the asteroid entering Earth's atmosphere, turning orange, sizzling.

You know what comes next: the impact in the Gulf of Mexico, earthquakes around the world, forests in the Northern Hemisphere spontaneously bursting into flames, the sky blackened for months by soot. The dinosaurs, and many other creatures, wiped out. A sad day, indeed. This Pixar offering, apparently, is darker than most of their movies, a tragedy, ending with the demise of the great reptiles.

Or maybe it isn't.

"But what if," the trailer asks, and then shows the asteroid streaking through the Cretaceous sky. Grazing behemoths—sauropods, duck-billed dinosaurs—look up momentarily, then go back to filling their cavernous bellies with leafy food. The asteroid flies by, a near miss instead of a fatal impact. Life goes on. The dinosaurs' salad days continue.

I know the answer to the question "What if?" The dinosaurs were at the peak of their reign sixty-six million years ago. They had dominated the world for more than one hundred million years. Sans asteroid, the dinos would have continued their global rule: *T. rex, Triceratops, Velociraptor, Ankylosaurus*—they all would have survived. New dinosaurs would have evolved, replacing the old ones. The ever-changing dinosaurian parade would have marched on. In all likelihood, the dinosaurs would still be walking the Earth today.

And who wouldn't be here today? We wouldn't, that's who. Even though we mammals evolved about 225 million years ago, almost exactly at the same time as the dinosaurs, for the first 160 million years of our existence, we didn't amount to much. The dinosaurs saw to that. Our furry forebears were an insignificant afterthought in the global biosphere, generally much smaller than the smallest dinosaur, active at night to avoid their reptilian overlords, scurrying in the underbrush, eating whatever scraps they could find. If you think of an opossum, you have a good idea of the looks and lifestyle of our Cretaceous kin, though most were probably even smaller.

It wasn't until the asteroid wiped out the dinosaurs that Team Mammal got its evolutionary opportunity—and we certainly took advantage of it, quickly proliferating to fill the empty ecosphere, transforming the last sixty-six million years into the Age of Mammals. But we owe all of that to the asteroid.

We—scientists and laypersons alike—once thought that the rise of mammals was inevitable, that we mammals are inherently superior to those reptilian brutes, thanks to our big brains and our internal combustion engines generating body heat. It took some time, so the idea went, but we eventually supplanted the dinosaurs, perhaps by eating their eggs into extinction or otherwise showing them who's who.

We now know this is nonsense. Mammals had bit parts in the Mesozoic evolutionary play. The dinos were doing just fine on that lovely day in 66 million BC, their dominance in no manner challenged by the vermin underfoot. Without the asteroid, life would have continued on its merry way, with reptilian intrigue and machinations, new species evolving, others going extinct, as they had for millions of years. There's little reason to think that we mammals would have emerged from the shadows to become major players in the ecosystem. The dinosaurs were already there, filling the ecological niches, using the resources—it was only after they were gone that we had our evolutionary turn.

No asteroid, no mass extinction, no mammal evolutionary flowering, no you and me. So, these first few moments of the movie trailer had me excited. Pixar had made a movie all about dinosaurs and how the world would have turned out differently if the asteroid had sailed on by. Forty-five seconds into the preview, I knew the movie was going to be a winner.

The trailer continued with a *T. rex* chasing a herd of plant eaters, causing them to stampede, a pell-mell rush of enormous herbivores, long-necked brontosaurs* and three-horned *Triceratops*, a typical day in the Mesozoic. But then I did a double take—some of those beasts looked more like hairy, big-horned bison than ceratopsians. And the next scene shows a brontosaur bounding along with something on its head—a human child!

If the asteroid was a near miss, what are mammals doing there? This is a Pixar movie, after all, so one expects a few liberties to be taken (dinosaurs speaking English, for example), but is there any scientific

---

* Dinosaur purists may note that the name *Brontosaurus* was long ago discarded, replaced for quirky scientific reasons with *Apatosaurus*. To those killjoy know-it-alls I respond, "Haha! Thanks to new scientific discoveries, the name *Brontosaurus* was resurrected in 2015."

evidence supporting the juxtaposition of *Brontosaurus*, bison, and baby? If the dinosaurs hadn't been wiped out, might mammals have diversified anyway, producing bison and—more importantly—us? Dinosaurs had kept mammals in their place—that place being tiny and in the underbrush—for millions of years. Is it possible that somehow, after all that time, mammals could have cut loose evolutionarily and prospered, even while the rule of the big reptiles continued?

There is one possibility, at least according to British paleontologist Simon Conway Morris. Dinosaurs, being reptiles, liked it hot. Their low metabolic rates did not produce much internal heat. When it was warm outside, that wasn't a problem—they could get their heat from the ambient environment, supplementing it when necessary by sitting in the Sun. The dinosaur dynasty was enabled by a long stretch of global warming, a time when much of the world was tropical, a good time to be a reptile.

But Conway Morris points out that the climate finally began to change about thirty-four million years ago. The world got cooler. Eventually, the ice ages came, glaciers expanded, much of the world became chilly. There's a reason you don't find reptiles in the far north and south today—it's too cold for them. Conway Morris suggests that even with dinosaurs still extant, this global cooldown would have sprung the mammals, kick-starting their evolutionary radiation. Dinosaurs would have had to retreat to the tropical equator, leaving the higher and mid-latitudes free, giving mammals their evolutionary chance at last.

Let's humor Conway Morris for the moment and assume his scenario is correct. Mammals start diversifying, occupying ecological niches long filled by dinosaurs, becoming bigger, more diverse. Maybe this Ice Age–enabled evolutionary diversification would have led to an Age of Mammals equally as magnificent and multifarious as the one the asteroid spawned.

But would it have been the same Age of Mammals? Would there be elephants and rhinos and tigers and aardvarks? Or would this alternative world have produced a very different ensemble of animals—species completely unrecognizable to us, dividing up the world's resources and filling its ecological niches, but in different ways than the creatures around us today? Or, to place the question closer to home, would we have evolved? Would there be humans to produce babies to sit atop Pixar's *Brontosaurus*?

Conway Morris responds with an emphatic "yes." To him and other scientists in his camp, evolution is deterministic, predictable, following the same course time after time. The reason, they argue, is that there are only so many ways to make a living in the world. To each problem posed by the environment, a single, optimal solution exists, leading natural selection to produce the same evolutionary outcomes over and over.

As evidence, they point to convergent evolution, the phenomenon that species independently evolve similar features. If there are limited ways to adapt to a given environmental circumstance, then we would expect species occupying similar environments to convergently evolve the same adaptations, and that's exactly what happens. There's a reason that dolphins and sharks look so much alike—they evolved the same body shape to move rapidly through the water in pursuit of prey. The eyes of octopuses and humans are nearly indistinguishable because the ancestors of both evolved very similar organs to detect and focus light. The list of evolutionary convergences goes on and on, as we shall soon see. Conway Morris and his colleagues see it as ubiquitous and inevitable, allowing us to predict how evolution would have unfolded, what a late-blooming mammal radiation might have looked like. Conway Morris concludes that "the rise of active, agile, and arboreal ape-like mammals, and ultimately a hominid-like form, would have been postponed, not cancelled . . . without the end-Cretaceous asteroid impact . . . the appearance of the hominids would have been delayed by approximately thirty million years." Pixar, in other words, was on solid ground in commingling babies and brontos.

But let's take this argument one step further. Even if the mammals forever stayed in the shadows, could a species like us have evolved from some other ancestral lineage? If convergence is so ineluctable, the push to particular solutions so unrelenting, there's no reason to think the rise of mammals was a necessary prerequisite. A big-brained, bipedal, highly social species with forward-facing eyes and forelimbs capable of manipulating objects could have evolved from some other ancestor. But if not from mammals, then descended from what?

Answering that question requires no more than switching from *The Good Dinosaur* to the bad dinosaur. Specifically, to *Velociraptor*, the villain of *Jurassic Park* (and, in an unexpected case of redemption twenty years later, the hero of *Jurassic World*). Talk about smarts! These wily reptiles worked as a team, outwitted the hardened safari hunter, and even figured out how to open doors with their three-fingered hands. And they were visually oriented and bipedal. Beginning to sound familiar?

With a few exceptions, *Jurassic Park*'s portrayal of *Velociraptor* was reasonably accurate.* Of course, we don't know how smart they were, but they did have large brains and some paleontologists have speculated that they may have been social, living in groups and coordinating their predatory attacks like lions or wolves. If you were looking for a jumping-off point for the evolution of a hominid-like animal, *Velociraptor* would seem like a good place to start.

And that's just where Canadian paleontologist Dale Russell began in the early 1980s. He studied a close relative of *Velociraptor*, another small theropod dinosaur named *Troodon* that also lived at the end of

---

* Although, in reality, the creature was based on the closely related dinosaur *Deinonychus*. One major difference between the movie and reality was that *Velociraptor* probably stood less than three feet tall. However, in an example of life imitating fiction, shortly after *JP* premiered, paleontologists described a larger cousin of *Velociraptor*, dubbed *Utahraptor*, which was about the size of the raptor in the movie.

the Cretaceous period. *Troodon* had the largest brain relative to its body weight of any dinosaur, a brain comparable in size to that of an armadillo or a guinea fowl. In other words, these reptiles were no geniuses, but they weren't completely clueless, either. Russell noted that over the course of hundreds of millions of years, animals have steadily evolved bigger brains. The fact that the largest dinosaur brain occurred in a species that lived at the end of their tenure suggested that dinosaurs, too, were following this evolutionary trend of increasing brain size through time. What would have happened, Russell asked, if the asteroid hadn't wiped them out? How would *Troodon*'s descendants have evolved if natural selection pushed them toward ever larger brains?

Russell went through a chain of logic to speculate what a modern-day descendant of *Troodon* would have looked like: larger brains require larger braincases; bigger braincases usually are associated with a shortening of the facial region; heavier heads are more easily balanced by placement directly on top of the body; this in turn favors an upright posture, which means that a tail is no longer needed as a counterweight to the no longer forward-leaning front half of the body. A few more assumptions about the best leg and ankle structure for walking upright and, voilà, what was termed inelegantly the "dinosauroid," a green, scaly creature with an uncanny resemblance to a human, right down to the butt cheeks and fingernails.

Remember, Russell did not set out to ask how a dinosaur could evolve into a hu-

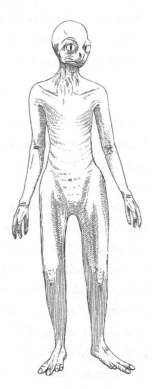

The dinosauroid

manoid. Rather, his goal was to think about how selection for increased brain size would lead to other anatomical changes. The end result of this project led to envisioning a creature strikingly similar to us, a reptilian humanoid.

Russell's evolutionary projection, though conjured years in advance, is consistent with Conway Morris' ideas that the evolution of hominid-like life-forms is inevitable. So consistent, in fact, that Conway Morris even appeared in a BBC documentary, sipping coffee at a café next to a dinosauroid reading a newspaper.

So Pixar had a couple of plot options. If the Cretaceous asteroid had, indeed, missed Earth, then according to Conway Morris and others, humans or something like us would have evolved one way or another. The only question was whether they would have been hairy, the result of delayed mammalian evolutionary diversification, or scaly, an outcome of natural selection on increased dinosaur brain size.

IT'S FUN TO THINK COUNTERFACTUALLY, to wonder what might have transpired if history had unfolded differently. But questions about the inevitability of humanoid evolution transcend speculation about Earth's history.

We now know there are a lot of planets in the universe that potentially could harbor life as we know it. These "habitable exoplanets" are neither too hot nor too cold and have liquid water on the surface. A recent study indicated that billions of such planets may exist in the Milky Way galaxy alone. The nearest may be only four light-years away.

Suppose life has evolved on some of these planets. What would it look like? Would the life-forms resemble those here? And what about intelligent life-forms, as smart as us, or even much smarter? How much, if at all, would they be like humans?

Quite a lot, if we believe what we see in the movies, and some well-renowned scientists agree. "If we ever succeed in communicating with conceptualizing beings in outer space," wrote the late biologist Robert Bieri, "they won't be spheres, pyramids, cubes, or pancakes. In all probability, they will look an awful lot like us." David Grinspoon, doyen of the emerging interdisciplinary field of astrobiology,* goes a step further: "when they [aliens] do finally land on the White House lawn, whatever walks or slithers down the gangplank may look strangely familiar." Not surprisingly, Conway Morris agrees, suggesting that "the constraints of evolution and the ubiquity of convergence make the emergence of something like ourselves a near-inevitability." But before exploring the scientific basis for these scientists' extraterrestrial predictions, let's return to Planet Earth.

MORE SPECIFICALLY, to southeast Africa. Darkness comes quickly in the Zambian woodlands. I'm a herpetologist—a lizard guy—so tracking nocturnal lions is not my day job, but I've come to Zambia for a little R & R prior to fieldwork in South Africa. Amazingly, lions can become accustomed to the presence of vehicles and will allow you to shadow them as they go on the prowl, and that's just what we're doing.

Off to the right there's a movement, something not too large approaching, unaware that it's on a collision course with a pride of lions. As it shuffles closer, its identity becomes clear—a crested porcupine, the sixty-pound rodent covered head to tail with pointy spines, some a foot and a half long. Its spines, of course, are for defense, for situations just like this, but they're not always effective. Lions have a counterstrategy, slipping a paw underneath the porcupine's body to flip it over, exposing the vulnerable belly. You can imagine the rest.

---

* Yes, that's a real scientific discipline devoted to the study of life elsewhere in the universe as well as the origin of life here on Earth.

There's a *Seinfeld* episode in which Jerry's watching a nature documentary on antelopes, and the lions attack, and Jerry's yelling, "Run, antelope, run! Use your speed. Get away!" And the next night, he's watching another nature flick, this time focused on lions, and they go for an antelope, and he's shouting, "Get the antelope; eat him; bite his head! Trap him; don't let him use his speed!" But even though tonight we've been following the lions, I'm rooting for the porcupine. Leave him alone and go after something your own size!

But, of course, they don't. One of the lionesses wanders over to the porcupine. He turns his backside to her, erects his spines, sort of like a cat arching its back and bristling its hair, and then he starts shaking the tail spines against each other, clackity-clack, clackity-clack.

And, amazingly, it works. After a moment, the lioness turns away and rejoins the pride, and the porcupine wanders off into the night.

At the end of the evening, I replayed the events in my head, mulling over my previous porcupine encounters. As well as Africa and Asia, porkies also occur throughout most of the New World. I've only seen the North American porcupine in the wild once, in a tree of all places—thirty feet up as I glided by on a ski lift. In the rainforests of Costa Rica, however, I've seen prehensile-tailed porcupines a number of times, again mostly in trees.

Certainly, there are differences among these species. The most obvious is size: the crested porcupine is twice the weight of its North American counterpart and thirty times that of the diminutive Rothschild's porcupine from Panama. The quills correspondingly vary in length—fourteen inches in the crested, four inches in the North American, shorter yet in the Rothschild's.* Some species have red noses, others brown; prehensile-tailed porcupines have no quills on their tails.

---

* These figures refer to the stiff, hardened quills that do the most damage. Thinner, more flexible quills are often substantially longer.

Two porcupines: The North American porcupine (left) and the African crested porcupine (right)

Yet, the differences pale in comparison to the similarities: not only possession of quills, but also a similar stocky body with short legs, small eyes, spiky hairdo. Given these similarities, I never questioned my assumption that porcupines were one happy evolutionary family, all descended from the same ancestral spiny ur-porcupine.

Imagine my surprise, then, when I learned that I had it all wrong. Despite their shared prickliness, New and Old World porcupines do not share a common evolutionary heritage. Rather than owing their pointy good looks to descent from a common, bristly ancestor, the two lineages have independently evolved their quills from different, unquilled rodent species. They are the result of convergent evolution.

I'M NOT THE FIRST PERSON in history to be fooled by convergence. In fact, I'm in pretty good company. Charles Darwin himself was bamboozled on his famous visit to the Galápagos Islands. There he discovered the small birds that now bear his name, the Darwin's finches. But Darwin did not realize that these bird species were all closely related to each other, descendants of a single ancestral finch that colonized the islands sometime in the past. Rather, he thought the species represented four groups with which he was familiar from home: true finches, grosbeaks, blackbirds, and wrens.

It was only when Darwin returned to London and turned his specimens over to the noted ornithologist John Gould that he learned his mistake. The species were not representatives of a diverse set of familiar types after all, but instead members of a single group of birds unique to the Galápagos—Darwin had been hoodwinked by convergent evolution. This revelation fit in with other findings from Darwin's voyage, all pointing in one direction, toward the "transmutability" of species. By the time that he revised his best-selling *Voyage of the Beagle* in 1845, the finch story intimated what was to come a decade later: "Seeing this gradation and diversity of structure in one small, intimately related group of birds, one might really fancy that from an original paucity of birds in this archipelago, one species had been taken and modified for different ends."

The broader implication of the story—that the finches had diversified on the Galápagos to mirror species using a variety of habitats elsewhere—was also not lost on Darwin. Although he didn't allude to convergent evolution in the *Voyage*, he clearly articulated the idea fourteen years later in the *Origin*: "in nearly the same way as two men have sometimes hit on the very same invention, so natural selection . . . has sometimes modified in very nearly the same manner two parts in two organic beings, which owe but little of their structure in common to inheritance from the same ancestor."

Darwin was not the only early naturalist so fooled by convergence. When Captain Cook landed in Botany Bay in 1770 on his first South Pacific voyage, the naturalist on the expedition, Joseph Banks, sent specimens and drawings of Australian birds back to England. This began a flood of material dispatched to the motherland by colonists and explorers over the next half century, revealing the existence of many new species.

The key figure in making sense of this profusion of new species was John Gould. At around the same time that he was consulting with Darwin on the finches, Gould decided to take up a comprehensive description of Australian birds. Quickly realizing that he needed to go to Australia to do the job right, he picked up and relocated Down Under, spending three years there and eventually producing a mammoth seven-volume series of paintings and descriptions.

But Gould, so right about Darwin's finches, turned out to be equally wrong about the evolutionary affinities of Australia's avifauna. Many Aussie birds are very similar in appearance and habit to species in Europe, such as wrens, warblers, babblers, flycatchers, robins, nuthatches, and others. As a result, Gould assigned the newly discovered Australian birds to the familiar Northern Hemisphere families.

Gould's error is understandable. Over the course of the next century and a half, many very knowledgeable ornithologists were equally deceived and treated these birds as colonial outposts, the result of a wave of invasions of Australia by many types of birds.

However, genetic studies starting in the 1980s showed that, in fact, most of the species are part of a large Australian bird radiation that evolved in situ. In other words, these Australian birds are closely related to each other; they are not members of many different Northern Hemisphere families, but convergent with them.*

The discovery of unexpected cases of convergent evolution continues to this day. Indeed, with the flood of genetic data now available for so many different species, our understanding of evolutionary relation-

---

* Not only was there no wave of bird colonizations into Australia, but the data indicate that a number of families, particularly songbirds, arose in Australia and dispersed from there to the rest of the world.

ships is advancing by leaps and bounds, producing a much firmer grasp on the evolutionary tree of life. One consequence is that we are increasingly finding new cases in which we had been misled by anatomical similarity, only now realizing that it results not from descent from a similar common ancestor, but from independent derivation.

How can we explain this rampant convergent evolution? There's a commonsense explanation, the one Darwin proposed. If species live in similar environments and face similar challenges to their survival and reproduction, then natural selection will lead to the evolution of similar traits: the existence of large seeds is a resource for birds, requiring big beaks to crack them open, and so similar, big-beaked birds evolve in numerous seedy locations; threatened by big cats, oversized rodents repeatedly evolve a spiny defense, as effective against lions in Africa as it is against pumas in the Americas.

In the last two decades, some biologists have extended this view to the cosmos. Here on Earth, species face the same challenges around the world and through time, and they evolve the same solutions. These scientists argue that the same physical challenges that occur here will also be faced by life-forms on similar planets and will lead to the same biological solutions. George McGhee, a paleontologist from Rutgers University, argues that there's only one way to build a fast-swimming aquatic organism, and that's why dolphins, sharks, tunas, and ichthyosaurs (extinct marine reptiles from the Age of Dinosaurs) all look alike.

Taking this a step further, he argues that "if any large, fast-swimming organisms exist in the oceans of Jupiter's moon Europa, swimming under the perpetual ice that covers their world, I predict with confidence that they will have streamlined, fusiform bodies . . . very similar to a porpoise, an ichthyosaur, a swordfish, or a shark." Conway Morris agrees, saying, "Certainly it's not the case that every Earth-like planet will have life let alone humanoids. But if you want a sophisticated plant it will look awfully like a flower. If you want a fly there's only a few ways you

Shark (top), ichthyosaur (middle), dolphin (bottom)

can do that. If you want to swim, like a shark, there's only a few ways you can do that. If you want to invent warm-bloodedness, like birds and mammals, there's only a few ways to do that."

NOT EVERYONE AGREES with this viewpoint. Let's go back to the movies to see why.

In the climactic scene of the classic 1946 film *It's a Wonderful Life*, George Bailey (played by Jimmy Stewart) despairs that his life has been a failure and wishes that he'd never been born. Clarence Odbody, George's guardian angel, then shows him how life in Bedford Falls would have been radically different—and much for the worse—if George had never existed: his brother dead; his friends and family unhappy, homeless, and institutionalized; a boat full of soldiers sunk; the town a den of iniquity. George realizes that his life has been worthwhile and abandons his suicidal plans, then subsequently is redeemed when the townspeople come to his rescue in appreciation of all his good deeds.

The American Film Institute in 2006 named *It's a Wonderful Life* the most inspirational movie of all time. Stephen Jay Gould, the famed paleontologist and evolutionary biologist, was among those inspired by it, but in a way different than most. To him, the movie was a parable for the evolutionary history of life, so much so that the title of his 1989 book, *Wonderful Life*, paid homage to the movie. In the book, Gould argued for the dominating importance of historical contingency in evolution. By contingency, he meant that the particular sequence of events critically determines the course of history: A leads to B, B to C, C to D, and so on. In a historically contingent world, if you alter A, you don't get D. If George Bailey is never born, events in New Bedford unfold differently.

Gould argued that life is full of George Bailey events—some major, most minor—but any of which could send life in a different direction.

Lightning strikes, falling trees, asteroid impacts, even the flip-of-a-coin determination of which genetic variant a mother passes on to her daughter—any of these could make a difference that would ramify through the eons. Like New Bedford without George Bailey, Gould wrote, "any replay [of the history of life] altered by an apparently insignificant jot or tittle at the outset, would have yielded an . . . outcome of entirely different form."

This view has important implications for understanding the diversity of life we see around us. If evolution is dominated by contingency, then there can be no predictability, no Conway Morrisian determinism. The end result is so influenced by contingencies that there is no way one could predict at the beginning what would happen at the end. Start over again, and a completely different result might unfold. Hitting home where it matters the most, Gould concluded, "Replay the tape [of life] a million times . . . and I doubt that anything like *Homo sapiens* would ever evolve again."

GOULD'S ARGUMENT, elegantly and persuasively made, resonates with us all. Who hasn't rued that "if I hadn't done X, then Y wouldn't have happened," where X could be anything minor (mispronouncing a name) or major (having a drink too many) and Y is something you wish hadn't taken place?

Still, sensible as the argument may be, what is the evidence? There's only one history of life. How can we test the repeatability of evolution? Gould proposed a thought experiment to address such questions. Replay the tape of life, he suggested, go back to the same starting conditions and see if the same result ensues. Such "gedankenexperiments," as the Germans call them, have a long pedigree in science and philosophy, and this one has been taken up by many and proven particularly fruitful.

Conway Morris and colleagues, of course, disagree with Gould's basic premise—changing an earlier event need not substantially alter the downstream outcome. They argue that the ubiquity of convergent evolution demonstrates the impotence of contingency, that in many cases more or less the same outcome would ensue regardless of the specific historical sequence of events.

The issue of convergence and evolutionary determinism had not yet been raised when Gould wrote *Wonderful Life*. However, in an exchange with Conway Morris published nine years later, Gould's response was simple: the importance of convergence is "overestimated," he said, and pointed to Australia as state's evidence number one.

Let's again consider Captain Cook's expedition to the Antipodes. Among the first animals they encountered was a kangaroo. Kangaroos are the major native plant eaters in Australia today. Functionally, they fill the same role as deer, bison, and myriad other herbivores in the rest of the world. And yet, as Gould (Stephen Jay, not John) noted, kangaroos haven't converged upon these other types of herbivores—even a toddler can tell that a kangaroo and a deer are different sorts of animals.

And then there's the koala, that lovable, bearish tree hugger that lives life in the slow lane, sleeping twenty hours a day as it detoxifies the eucalyptus leaves that comprise its diet (and that make its fur reek of menthol). Nothing like it exists anywhere else in the world, now or, according to the fossil record, ever.*

But when we're talking evolutionary one-offs, there's only one king. Venomous ankle spurs, luxurious pelt, the ability to detect the electrical discharges of their prey's muscles with electroreceptors on their snout. Powerful flat tail, webbed feet, lays eggs. Bill like a duck. The world's greatest animal, the duck-billed platypus, a mishmash of parts

---

* Although, curiously, the fingertips of koalas are covered with ridges and whorls so similar to our own that experts are hard-pressed to distinguish between fingerprints of koalas and humans.

The duck-billed platypus

borrowed from throughout the animal kingdom. An animal so confused that when the first specimens arrived in England at the end of the eighteenth century, shipped from Sydney across the Indian Ocean, scientists searched for hours in vain to locate the stitches by which crafty Chinese merchants must have assembled their hoax.

These examples have come from Down Under, but evolutionary one-offs occur everywhere. Giraffes, elephants, penguins, chameleons—these are all species exquisitely adapted to their specific ecological niches, with no evolutionary facsimile now or in the past (note that an "evolutionary one-off" is not necessarily a single species. For example, there are three living species of elephant, and many more that occurred in the past, like mastodons and mammoths. However, all elephant species are descended from a single ancestral elephant. That is

why elephants can be considered evolutionarily unique—the probosci-
dean way of life only evolved a single time).

CONVERGENT EVOLUTION is a scientific phenomenon, and you'd think
that science should have been able to settle the question of its ubiquity
by now. But the problem is that figuring out what happened in the past
is not easy. We are taught in grade school about the scientific method,
how observations lead to the formulation of a hypothesis that is then
tested with a decisive experiment in the laboratory. That formulation in
a very simplistic way captures the operation of mechanistically oriented
sciences—that is, the sciences involved in understanding how something like a cell or an atom works. Think that a particular gene is important in producing a particular trait? Use molecular biology wizardry
to disable the gene and see if the trait still develops.

But evolutionary biology is a historical science. Like astronomers
and geologists, we evolutionary biologists try to figure out what happened in the past. And like historians, we are bedeviled by the asymmetry of time's arrow—we can't go back in time to see what happened.
Moreover, evolution occurs notoriously slowly, seemingly making it
impossible to watch it as it occurs.

Stephen Jay Gould laid out the experiment we'd like to do: replay
evolution time and time again, and see how sensitive the outcome is to
various experimental perturbations. But we call such ideas thought experiments for a reason—in the real world, they can't be conducted. Or
so we used to think.

It turns out that Darwin and a century of biologists following him
were wrong in one key respect: evolution does not always plod along at
a snail's pace. When natural selection is strong—as occurs when conditions change—evolution can rip along at light speed (I'll tell the story

of how we came to realize that evolution is as much a hare as a tortoise in Chapter Four).

The reality of rapid evolution allows us to go beyond simply observing whether and how species respond. In a development that would have astonished Darwin, researchers are creating their own evolution experiments, altering conditions in a controlled and statistically designed way. Just like lab biologists, we can test evolutionary mechanisms, but out in nature, in real populations. Researchers are placing light- and dark-colored mice in half-acre-sized cages in the Nebraska sand dunes, moving guppies in Trinidad from stream pools with predators to those without, and switching walking stick insects from one habitat to another.

I've conducted some of these experiments myself, testing hypotheses about why small lizards in the Bahamas evolve longer or shorter legs. I know what you're thinking, but my colleagues and I are willing to sacrifice for science. It's a dirty job, hanging out on beautiful, windswept islands surrounded by ocean, but someone's got to do it, and we're the ones. I'll go into much greater detail in Chapter Six, but for now suffice to say that if you go back to the Bahamas year after year and measure the legs of thousands of lizards with a portable x-ray machine, you'll see that lizard populations can evolve rapidly. Moreover, if you experimentally alter the conditions the lizards experience, causing them to change their habitat use, populations on those islands will evolve quickly and in predictable directions.

Although evolution experiments in nature are still in their infancy, laboratory scientists have been conducting such work for decades. These studies trade the realism of nature for the hyper-precision of the lab, providing exquisite control over conditions experienced by the evolving populations. Moreover, the shorter life span of lab organisms, particularly microbes, means that these studies can be longer-term, encompassing more generations and creating more opportunity for evolution to occur.

One laboratory experiment has been following microbial evolution for more than a quarter century, studying the extent to which twelve populations evolve in the same way.

I OFTEN COMPARE EVOLUTIONARY BIOLOGY to a detective story, a whodunit. A crime has been committed—or in this case, something has evolved—and we want to know what happened. If we had a time machine, we could go back and watch for ourselves. If we could replay the tape, we'd just set things up like they were back then and start it again.

But neither of these is possible (with one important exception I'll get to in Chapter Nine). Instead, we're left with a bunch of clues, and, like Sherlock Holmes, we have to figure it out as best we can. We can see the patterns of evolutionary history, the species that occur today and the fossils of what existed in the past, allowing us to assess the extent to which evolution has repeatedly produced the same outcome. And we can study the evolutionary process as it operates today. By conducting experiments, we can see how repeatable and predictable evolution is: If you start at the same point, will you always end up with the same outcome? And if you start at different points, but select in the same way, will you converge on the same result? So even though we can't replay the tape, we can study evolutionary pattern and process. By putting the two together, scientists are now well on the way to understanding evolutionary repeatability.

This book, then, is about the extent to which life repeats itself, the result of species evolving similar adaptations in response to similar environmental circumstances. More loftily stated, it is about determinism, whether natural selection inevitably produces the same evolutionary outcomes or whether the particular events a lineage experiences—the contingencies of history—affect the end result.

At the same time, this is a book about how scientists study these topics, how tools from DNA sequencing to fieldwork in remote corners of the world are synthesized to understand the evolutionary origin of life around us. And it's also about how science itself evolves, how new ideas are born and how research programs develop to test them. In particular, I'll focus on the rise of experimental methods to studying evolution, an approach that was inconceivable for more than a century after Darwin's time.

The book will be full of scientists and their research, in pristine lab and woolly nature, but the topic is of more than academic interest. Evolution is occurring all around us today and has consequences beyond informing our understanding of arcane debates. Most notable are the direct evolutionary battles between humans and our commensals. On the one hand, nature is fighting back against our efforts to control it. We consider some species to be pests because they have the audacity to use the resources we wish to reserve for ourselves. Weeds invading our fields, rats eating our grain, insects devastating our crops. We deploy an armory of chemical—and, increasingly, genetic—weapons to control them, but they quickly evolve ways around them.

Seven billion and counting, sometimes we are the resource being exploited. Malaria, HIV, hantavirus, influenza—to microorganisms, our bodies are like any other crop and they are evolving to take advantage of us. We, in turn, combat them as we do crop pests, with chemicals, and they rapidly evolve resistance.

This is where the debate between contingency and determinism becomes personal. If we can predict not only when rapid evolution will occur, but what form it will take, we will be able to derive general principles and thus be better positioned to respond effectively. But if each case of rapid evolution is contingent on the specific circumstances, then we'll have to start from scratch each time we face a new weed, pest, or disease, figuring out how our evolutionary foe is adapting and what we can do about it.

DEBATE ABOUT CONTINGENCY versus determinism affects us in another, more ethereal way. Humans are no less subject to convergent evolution than other species. Our ability to drink milk as adults, for example, is unique among animals; it was, of course, irrelevant until we domesticated livestock in the last few thousand years, and since then has evolved convergently in several pastoral societies around the world. Skin color, so important in the course of human history, is also the result of convergent evolution, as is the ability to survive at high elevations, and many other traits.

The human species itself, of course, is not convergent. We are one of the singletons, lacking an evolutionary duplicate. Does our understanding of evolutionary determinism have anything to say about how we evolved, or why? If we hadn't come along, would some other lineage have taken our place, and would that species have ended up much like us, perhaps so much so that someone—something—else would have been writing this very book, albeit with scaly, three-fingered hands? And if not here, perhaps on the moons of Jupiter or xh3-9?

But once again, I get ahead of myself. Let's return once more to Earth and see just how pervasive convergent evolution is on our own planet.

Part One

NATURE'S DOPPELGÄNGERS

chapter one

Evolutionary Déjà Vu

Picture a whale swimming through the ocean: streamlined body, flippers, a small fin on the back, its tail undulating up and down. Given this piscine countenance, who could fault the ancient Greeks for thinking whales were a type of fish? That view persisted for millennia until Carl Linnaeus set things straight 250 years ago, recognizing the leviathans as mammals on account of their live birth, mammary glands, and other traits.* The Greeks had been tricked by convergent evolution.

We've come a long way since pre-Linnaean scientists. We certainly know a lot more about evolution than they did, and our enhanced understanding of anatomy and the evolutionary relationships of species has identified countless cases of convergent evolution. Nonetheless, our list is far from complete. As new data from molecular biology floods in, we're discovering time and time again that we've been misled, just as

---

* We now know that not all mammals give birth to live young. Platypuses and echidnas—grouped together as monotreme mammals—lay eggs. The production of milk and possession of hair are the two most obvious features of all mammals (though some mammals, like whales, have just a few whiskers).

the Greeks were, and that species we thought similar due to inheritance from a common ancestor instead have independently converged upon the same traits.

Let me provide two recent examples. By some measures, sea snakes are among the most deadly serpents, the venom of some species, drop for drop, as lethal as that of any ophidian. Fortunately, most sea snakes rarely bite even when handled. Not so, however, for the beaked sea snake, which defends itself fiercely and accounts globally for ninety percent of sea snake–caused human fatalities. Named for the tip of its snout, which overhangs the lower jaw, the species can be very common locally and has an enormous geographic distribution, from the Gulf of Arabia to Sri Lanka, Southeast Asia, and down to Australia and New Guinea, making it one of the most widely distributed snake species in the world.

Or so it was thought. In 2013, a team of Sri Lankan, Indonesian, and Australian scientists reported that they had conducted routine genetic comparisons among populations of the species and had gotten a most decidedly non-routine result. Even though the populations exhibit only minor anatomical differences across the species' range, they were highly divergent genetically. In particular, Australian populations of the beaked sea snake were most genetically similar to other Australian sea snake species rather than to Asian populations of their own species; similarly, Asian beaked snake populations allied most closely with other Asian species. In other words, there is not one species of beaked sea snake, but two. And the traits that define the species, not only its beak, coloration, and general appearance, but also its nasty disposition, have evolved convergently, so much so that distant relatives on opposite sides of the Indian Ocean were considered to be members of the same species.

And now for an example more familiar to those who have never seen a sea snake. As a lad, I was pure of mind and body, late to take up the joys of stimulants and debauchery. One day early in my adulthood, I was visiting with a friend. She offered me some tea. I was not a tea drinker, but I wanted to appear worldly and agreeable, so I accepted. Soon, I began to feel funny. My body was tingling, my hands shaking, my heart racing. I thought I might be having a heart attack. But, I reasoned, I was too young for that, plus coronaries weren't supposed to be so energizing. I can't remember exactly how I coolly queried my host, no doubt suavely admitting that I felt a little bit unusual, but she quickly explained that I was drinking a particularly invigorating brand of tea, the closest thing back then to Red Bull. Now as an adult, I get going in the morning with a cup of java, but I religiously avoid the stuff after four in the afternoon. Coffee any later than that and I'll be up all night.

Maybe you're different, but for me, life seems to involve continually relearning the same lessons. And so there I was, one recent night in the Pantanal of Brazil, tossing and turning, unable to drift off despite a hard day and a heavy meal. "Why can't I sleep?" I wondered, as my mind raced from one thought to the next. And then an epiphany. That unfamiliar, fruity soft drink at dinner. I was thirsty and had two cans. It was fizzy and vaguely tasted like apple juice. What was it?

Some quick-fingered sleuthing on the keyboard tracked down the name of the soda, Guaraná Antarctica, and what it is made from, the guaraná plant, a large-leafed climbing plant related to maples that hails from the Amazon rainforest. And guess what guaraná seeds are loaded with. The same compound that is in coffee and tea, Pepsi and Mountain Dew, and the chocolate in Hostess Ding Dongs. A purine alkaloid, the chemical 1,3,7-trimethylpurine-2,6-dione. Molecular formula: $C_8H_{10}N_4O_2$.

Caffeine.

Despite my awareness of its many delivery vehicles (Pepsi, tea, energy drinks), I had never given much thought to where the caffeine itself comes from. Coffee and tea come from eponymous plants; cola sodas, at least originally, came from the nut of the kola tree; chocolate from cacao; Guaraná Antarctica from the seeds of the guaraná plant (twice as caffeine-laden as coffee beans). All of these plants produce caffeine. And not different varieties, but the exact same molecule. Caffeine is caffeine, regardless of where it comes from. One molecule, many sources.

Because I'm an evolutionary biologist, my curiosity should have been piqued by the many different plants that produce caffeine, stimulating me to wonder whether these plants are all closely related or whether caffeine production has evolved convergently many times. But I was asleep on the job and the thought never occurred to me.

Fortunately, some more inquisitive botanists decided to investigate just this question. In a paper published in 2014, an international team of researchers used genetic data in a two-pronged approach to demonstrate that caffeine production has evolved independently in these plants. Prong one compared the DNA of many plant species to build an evolutionary tree of relationships among the caffeinaceous species (focusing only on three: coffee, tea, and cacao). Such evolutionary trees—"phylogeny" is the technical term—are like family genealogies. Closely related species occur near each other and can trace their ancestry to a recent common ancestor, just as brothers and sisters trace their lineage a short distance back to their parents. Distant relatives, like fourth cousins twice removed, occur on relatively distant branches of the phylogeny, and you have to work your way deeper down in the tree—further back in evolutionary time—to find their most recent common ancestor.

The team's phylogeny showed that coffee, tea, and cacao plants occur on different branches of the evolutionary tree—they are not closely related to each other. Rather, cacao is more closely related to maple and eucalyptus trees than it is to either tea or coffee. Similarly, coffee is de-

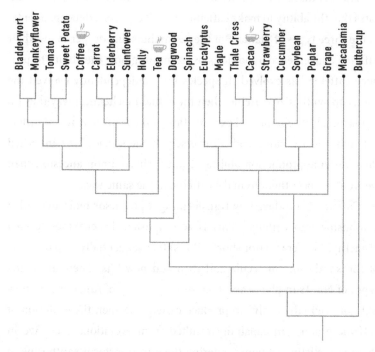

A phylogeny illustrating the evolutionary relationships of selected eudicot plants (plants with a specific type of pollen, constituting more than half of all plant species). Species that share a common ancestor are more closely related to each other than to species not descended from that ancestor. The steaming mugs represent species that produce caffeine. Because these three species are not closely related, the most likely interpretation is that caffeine evolved independently in each of the groups (an alternative possibility is that caffeine production was the ancestral state and was independently lost many, many times, but that scenario requires many more evolutionary changes and thus is less likely).

scended from an ancestor that also gave rise to potatoes and tomatoes, but not tea or cacao. Tea is on its own evolutionary branch, distant from all the other species in the study. Put another way, we have to go deep in the phylogeny, way back in evolutionary time, to find the ancestor that gave rise to tea, cacao, and coffee.

The fact that caffeine-producing species are not closely related indicates that the ability to make caffeine most likely evolved independently in the three types of plants. But the researchers dug deeper to test their caffeine convergence hypothesis by examining how the ability to produce caffeine has evolved. If species have independently evolved the ability to synthesize caffeine, then the actual biochemical way they do so may not be the same and examination of the DNA may reveal different routes to the same end. Conversely, if the species have inherited their caffeine-production abilities from their common ancestor, then we would expect them to make caffeine in the same way.

Caffeine is produced by transforming a precursor molecule called xanthosine into caffeine. This is accomplished by enzymes termed N-methyltransferases (for short, NMTs) that sequentially snip off parts of the xanthosine molecule and then add new bits. There are many types of NMTs in plants and they serve a variety of functions, so they did not originally evolve to produce caffeine. Rather, the evolution of caffeine-production capability resulted from evolutionary change in these pre-existing enzymes, altering them to transform xanthosine to caffeine.

By examining the genome of the different species, the researchers isolated the DNA of the different NMTs and discovered that the NMTs that were modified in coffee were different from the ones modified in tea and cacao. Thus, the evolutionary routes to caffeine production were different—convergence occurred through different evolutionary paths.

EVOLUTIONARY BIOLOGY is unlike many sciences in that its basic findings about the history of life cannot be derived from first principles. It is not a deductive science. You can't go to the chalkboard and derive the formula for a platypus. Rather, it is an inductive science in which general principles emerge from the accumulation of many case studies. These piles of research allow us to distinguish what occurs regularly from what happens only rarely. Put another way, evolution occurs in many different ways—just about anything plausible you can imagine has evolved somewhere at some time in some species. Given enough time, even the improbable will occur eventually. As the mathematician Ian Malcolm said in *Jurassic Park*, "Life finds a way." Thus, to understand the major patterns in the evolution of life, we ask not "What can happen?" but "What usually happens?"

And so it is with evolutionary convergence. The standard wisdom is that convergent evolution happens, but is not necessarily the expectation. Scientific papers routinely use words like "stunning," "striking," and "unexpected" to report its occurrence. News stories echo this sentiment, treating the publication of each additional example as if it were amazing and unanticipated.

But all that is changing. In recent years, a cadre of scientists has taken the opposite view, arguing that convergence *is* the expectation, that it is pervasive, and that we should not be surprised to discover that multiple species, often distantly related, have evolved the same feature to adapt to similar environmental circumstances. From this perceived ubiquity, the scientists draw a broader conclusion: evolution is deterministic, driven by natural selection to repeatedly evolve the same adaptive solutions to problems posed by the environment. In this view, the contingencies of history play a minor role, their effects erased by the predictable push of natural selection.

AT THE FOREFRONT of this movement has been Simon Conway Morris. Mild-mannered and self-deprecating, the University of Cambridge paleontologist doesn't seem like the sort to rock the boat. But under this unassuming exterior lies a sharp-witted pugilist who has spearheaded a radical reconsideration of the role of replication in the evolutionary pageant.

That Conway Morris should be an evangelist for convergent evolution and a fierce critic of Stephen Jay Gould might at first be surprising. As a young whippersnapper at the University of Cambridge, he made a name for himself with his doctoral research on the bizarre animals of the fabled Burgess Shale geological formation of the Canadian Rockies. But that research focused on a phenomenon that was seemingly the antithesis of convergent evolution.

The Burgess Shale formed around 511 million years ago, during the Cambrian period, when animal life as we know it was just emerging. Before then, life-forms were simpler, usually more or less flat, and unfamiliar. How life transitioned from this alien world to the ancestors of today is still debated, but it happened quickly and prolifically, giving rise to the Cambrian Explosion, when most of life's familiar kinds of animals—mollusks, echinoderms, crustaceans, vertebrates—first appeared in the fossil record in a geologically short period of time.

But it wasn't just the ancestors of today's fauna that appeared then. When the Burgess Shale fossils were first discovered in the early twentieth century by Charles Walcott, a paleontologist who was at that time the director of the Smithsonian Institution, they were all identified as belonging to well-known taxonomic groups—mollusks, crustaceans, worms, and so on. But when Conway Morris went back to reexamine the specimens a half century later, he found that many of these Cambrian species were paleontological weirdos, with no clear affinity to any recognized taxon (a taxon is an evolutionary group, such as fish or mollusks; multiple groups are taxa; the word "taxon" can apply to any evolutionary level, from species or genus to kingdom). Walcott, perhaps too distracted by administrative responsibilities or simply too narrow-minded to see them for what they were, had pigeonholed many of the Burgess Shale fossils into existing taxonomic categories despite their many oddities.

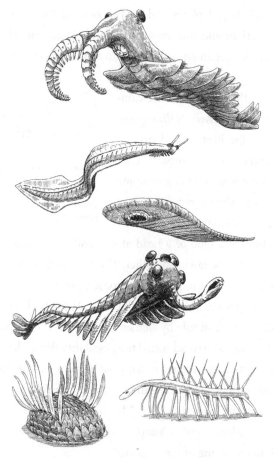

A sampling of the occupants of the Burgess Shale ecosystem 511 million years ago. From top to bottom: Anomalocaris, Pikaia, Odontogriphus, Opabinia, Wiwaxia *(left), and* Hallucigenia *(right).*

The term "weirdo" is not standard scientific parlance, but it gives a good sense of how peculiar they are. This realization came to Conway Morris as he painstakingly examined the tens of thousands of specimens collected by Walcott and residing in the musty drawers of the Smithsonian and other museums. Consider, for example, *Wiwaxia*, which looks like a pinecone lying on its side, lacquered in overlapping, oval-shaped plates. Add a flat bottom like a snail to glide along the seafloor and two rows of tall pointy spikes running down its back and you've got an animal similar to something out of a *Futurama* episode.

And then there's the creature Conway Morris christened *Hallucigenia*, referring "to the bizarre and dream-like appearance of the animal." "Cartoonish" is the word that comes to my mind. Conway Morris' reconstruction shows it to be a long, pencil-like tube

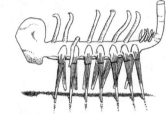

Simon Conway Morris' original reconstruction of Hallucigenia

with an ill-defined blob of a head at one end and a short, upturned, Scottish-terrier-style tail at the other. The tube body is festooned with seven pairs of pointy, unjointed stilts for legs, matched above by seven soft, squiggly tubes running down the back. At the back end, two rows of three short tubes sit side by side on the tail (assuming that part was the tail—Conway Morris admitted the possibility that he had the direction of the animal reversed such that what he thought was the head was actually the tail and vice versa—in his defense, the fossil was squashed flat and not of the highest quality).* In the paper announcing the species, Conway Morris put it simply, *Hallucigenia* "cannot readily be compared to any living or fossil animal."

And these were not the only oddities—indeed, the Burgess Shale was inhabited by a veritable bestiary of the bizarre: *Opabinia*, whose five eyes and long, claw-tipped hose on the front of its head provoked an audience of scientists to burst into laughter when it was unveiled for

---

* Thanks to new, better-preserved specimens, we now know that Conway Morris, through no fault of his own, reconstructed *Hallucigenia* upside down and backward. Those stilt-like legs were actually spines on the back, and the seven squiggly tubes on top were actually the legs, the second row of seven legs not detectable in the fossils he examined. In addition, better-preserved fossils showed the tail end to be the head and vice versa.

the first time; *Anomalocaris*, various parts of which were originally described as three different species until scientists realized they were all part of the same animal; *Odontogriphus*, a long, flat, soft animal that resembled a floating Band-Aid with a circular mouth on the underside of the front end. And the list goes on and on.

Stephen Jay Gould made the Burgess Shale oddballs famous in *Wonderful Life*. Subtitled *The Burgess Shale and the Nature of History*, the book was a detailed examination of the Burgess fossils and what they can tell us about evolution. And *Wiwaxia* and company weren't the only ones Gould made famous: the primary scientific hero of the book was none other than Simon Conway Morris, who had done so much to document how the Burgess Shale fauna was populated by so many unique life-forms so unlike anything previously known (Gould also lauded Conway Morris' doctoral advisor, Harry Whittington, and his fellow graduate student, now Yale professor, Derek Briggs).

In *Wonderful Life*, Gould dwelled on the outlandish anatomy of the Burgess Shale's inhabitants, arguing that the Cambrian fauna was the most diverse in Earth's history, pointing out the many anatomical forms that appeared and subsequently disappeared, nothing like them ever to be seen again. Gould speculated about why some of those ancient beasts survived and prospered, giving rise to today's diversity, while others perished. Were the survivors superior in some sense, destined to thrive, while the losers had evolved an inferior design? Or was it a matter of luck that some made it and others didn't? Gould concluded that there was no reason to believe that the survivors were necessarily adaptively superior to those that perished. Rather, it was happenstance, a lottery, that led some to survive and others to disappear. If life's narrative had been a little bit different, the tape replayed in a slightly different manner, he suggested, the world likely would be populated by a very different roster today.

Gould concluded *Wonderful Life* by focusing on one fossil in particular. *Pikaia* was a small animal that looked somewhat like a worm squished in a vise, vertically flattened and with no distinct head. This unprepossessing creature was the earliest known representative of the chordates, the evolutionary group containing vertebrate animals (that is, those with a backbone, like frogs, sharks, gorillas, and you and me).

In all respects, *Pikaia* was not a major Burgess player. Judged by the number of fossils discovered, it wasn't very abundant, and its size and shape were not very impressive. Amidst the great variety of species present then, a Cambrian observer would have been unlikely to pick out this species as a herald of great things to come. What if it was only a matter of luck that *Pikaia* survived while so many others died off? Replay the tape again, and *Pikaia* might not have made it. And if *Pikaia*'s line had perished, who would be ruling the world today? Not chordates, because we wouldn't be here.*

The argument for contingency was fashioned by Gould, but his lines of evidence, even some of his major supporting arguments, were pulled straight from the pages of Conway Morris' papers, as Gould repeatedly, exaltingly, emphasized.† Gould even suggested that for their accomplishments, Conway Morris and his two collaborators deserved a Nobel Prize in paleontology—if only there were such a thing.

But something funny happened on the way to Stockholm. Conway Morris, who had so emphasized the distinctiveness of so many of these

---

* Although still good for rhetorical effect, Gould's argument is diminished today by the fact that several other chordates have been discovered from the Burgess Shale and other similarly aged deposits. Consequently, even if *Pikaia* had perished, the entire chordate lineage would not have gone with it.

† For example, Conway Morris wrote, "If the clock was turned back so metazoan diversification was allowed to re-run across the Precambrian-Cambrian boundary, it seems possible that the successful bodyplans emerging from this initial burst of evolution may have included wiwaxiids rather than molluscs." And "a hypothetical observer in the Cambrian would presumably have had no means of predicting which of the early metazoans were destined for phylogenetic success as established bodyplans and which were doomed to extinction."

fossils, came to see the world in a different light. Rather than dwelling on the evolutionary uniqueness of so much of its fauna, Conway Morris concluded *The Crucible of Creation*, his own book about the Burgess Shale published in 1998, with a discussion of the importance and ubiquity of evolutionary convergence.

At face value, this reading of the rock record seems illogical—how do you go from celebrating the diversity of idiosyncratic, never-again-seen anatomies to seeing evidence for evolutionary replication everywhere? Conway Morris himself isn't sure, as he told me several years ago over lunch at St. John's College in Cambridge.

To some extent, he said, the explanation lies in new discoveries in the nearly three decades since *Wonderful Life*. Whereas a lot of the Burgess Shale species previously could not be associated with any known taxonomic group, newly unearthed fossils and detailed examinations have shown that many can now be assigned to recognized taxa. *Hallucigenia*, for example, appears to be related to modern-day velvet worms, an obscure, mostly tropical group of small animals that look like a cross between a centipede and a caterpillar; *Wiwaxia* is now thought by many to be related to mollusks.

So many of the Burgess Shale eccentrics are not so taxonomically iconoclastic after all. In addition, some analyses have compared the anatomical diversity of the Burgess Shale fossils to their modern counterparts, and have concluded—though this point is hotly contested—that the Burgess Shale fauna was no more diverse than living species are today.

These findings force a reconsideration of the Burgess Shale. Gould, following Conway Morris and his colleagues, had painted the Cambrian as a time of unparalleled anatomical diversity, occupied by a tremendous number of different types of organisms, most of which died out shortly thereafter. Ever since then, argued Gould, we have lived with a much-restricted range of anatomical design, all descended from the relatively few types that survived past the Cambrian.

Most researchers consider that the tide has turned on this viewpoint. The anatomical disparity was not so exceptional in the Cambrian, and the many forms that lived then do not represent failed evolutionary experiments that left no descendants today, but rather early relatives of today's surviving groups. Indeed, this was the thesis of Conway Morris' book, which in many respects was a sharply worded rejoinder to *Wonderful Life.*

Still, it's not clear why Conway Morris went from detailing Cambrian curiosities to cataloging convergence. The rescue of the Burgess species from taxonomic no-man's-land doesn't lessen their anatomical distinctiveness. Even if *Hallucigenia* is in the velvet worm lineage, for example, it is still anatomically unlike anything else that has ever evolved—these clarified phylogenetic relationships don't really make a case for convergent evolution.

One possible explanation for Conway Morris' about-face is that he was influenced by the direction the field was taking. In the mid-1980s, evolutionary biologists were increasingly employing the "comparative method," the idea that by comparing different taxa and looking for repeated patterns, one can find evidence for the operation of natural selection. Although this work was far from Conway Morris' research area, perhaps this emphasis on the importance of convergence shaped his thinking (although nothing he has said or written suggests this possibility).

We could also try our hand at psychoanalysis. Many are surprised at how critical Conway Morris has been of Gould, particularly given Gould's lionizing treatment of Conway Morris in *Wonderful Life.* One colleague proposed that Gould's views on the haphazard nature of evolution conflicted with Conway Morris' spiritual views. Another suggested that Conway Morris was embarrassed that Gould had publicly— in a best-seller!—trumpeted Conway Morris' earlier taxonomic views that subsequently had turned out to be mistaken. Whatever the cause of his antipathy, Conway Morris may have been primed to find ways to oppose Gould. In our conversation, Conway Morris recalled reading *Bully for Brontosaurus*, a collection of Gould's essays, and noting a number of cases of convergence that Gould failed to remark upon. Perhaps this was all it took to get Conway Morris thinking about convergence's evolutionary significance.

In any case, with the enthusiasm of a convert, Conway Morris has become the leading proponent of the view that convergent evolution is the dominant story behind life's diversity. "Evolutionary convergence is completely ubiquitous," he has said. "Wherever you look you see it." Consequently, he concludes, "Rerun the tape of life as often as you like, and the end result will be much the same."

UBIQUITY IS IN the eye of the beholder, but it would be hard to argue that convergence isn't common. In some cases, two species have independently evolved to be similar in one respect, such as the length of their tail, the color of their ears, the structure of their kidneys, even their mating dance. In more dramatic cases, species can be convergent in many different aspects of their phenotype, so much so that the two may appear to be indistinguishable, such as the two species of beaked sea snakes (the term "phenotype" refers to all the characteristics of an organism, everything from external anatomy to physiology to behavior).

Let's start by examining a few of the many different types of phenotypic traits that have evolved convergently. In recent years, scientists have identified convergence in almost any type of trait you might imagine. For example, many types of lizards have independently evolved flaps of skin under their necks that can be pulled out quickly like a semaphore to signal to mates or competitors; similarly, many birds have evolved colorful patches on their wings or breasts that are displayed prominently in social interactions. The natural world is full of examples of this sort: similar features, used in similar contexts, evolving multiple times in similar types of plants and animals.

Particularly impressive are traits that are convergent at an exquisitely detailed level, between species not at all closely related, from different parts of the tree of life. Here's a classic example: check out the eyeball pictured below.

If you remember your anatomy from whenever you learned it in school, that's a pretty typical peeper: could be a cow, or a human, or a cat, or even a lizard—the eyeballs of most vertebrates are pretty similar in basic structure. But that's no vertebrate's orb— that one belongs to an octopus! That's right—octopuses have eyeballs that are nearly

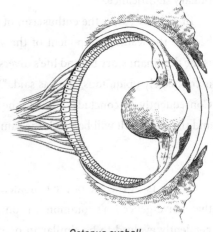

Octopus eyeball

identical to yours and mine, even though our most recent shared ancestor, which swam the Earth more than 550 million years ago, had no eyes to speak of.*

Or how about this one? Everyone knows praying mantises: big bug

---

* In fact, in some ways the octopus eyeball is better than yours or mine. In vertebrate eyes, the nerves attach to the front end of each of the retina's photoreceptors within the eye itself, meaning not only that light has to pass through the nerves to get to the receptors, but also that when the nerves bundle together to exit the eyeball, they create a photoreceptor-free area in the retina, creating our famous blind spot. By contrast, the design of the octopus eyeball is much more sensible, with the nerves attached to the back end of the photoreceptors, where they neither impede incoming light nor cause a visual obstruction when they exit out of the back of the eyeball. If, in fact, evolution did not occur and life was created by an intelligent designer, that designer apparently practiced on us before creating the better-designed eye of the octopus.

eyes, long neck, arms folded in prayer. But they're not actually as devout as they appear—their supplicating posture in reality is a mousetrap primed to fire, its lightning-quick strike snagging prey between its spine-encrusted forearm segments (as if we could catch lunch by quickly rotating our hands downward, pinning something between the palm of our hands and our forearms ... if our palms were covered with spines and half as long as our forearms).

But mantises aren't the only quick-draw artists in town. There's another type of insect called a mantidfly, which has nearly identical forearms for capturing prey with the same Superman-fast action. And the

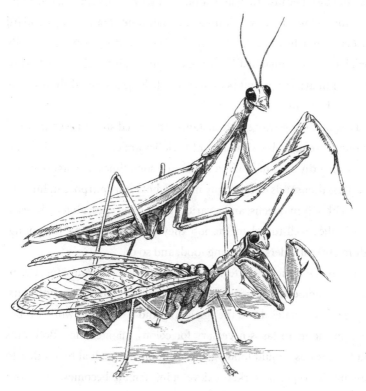

Praying mantis (top) and mantidfly (bottom)

similarity doesn't end there—the mantidfly's long neck and bulging eyes are so similar that its front half is a virtual mantis carbon copy, even though the two insects are separated by hundreds of millions of years of insect evolution (by contrast, the back half of the mantidfly looks more like that of its close relative, the lacewing).

Convergent evolution isn't limited to anatomy, of course. Species can converge in any attribute of their biology, from genes to behavior. There are many such examples, but some of my favorites come from the lowly ants and termites.

Most people assume that ants and termites must be closely related because you call the exterminator if you have a problem with either one, and also because they look alike. But if you pull out a magnifying glass for a closer look, you'll discover that, other than being standard insects with a head, a thorax, an abdomen, and six legs, they really don't look very similar at all. They also aren't at all closely related. Ants' nearest kin are wasps and bees; termites belong to—of all things—the cockroach family.

Despite their phylogenetic distance, the social structure of ants and termites is remarkably similar. Ant societies are characterized by a sophisticated division of labor: a queen (or sometimes queens) that lays countless thousands of eggs; tiny males whose only purpose in life is to mate with virgin queens; and a variety of worker types, all female, each with bodies well-tuned to the job they perform—caring for young, fighting off intruders, collecting food, and so on.

The social structure of termites is very similar. Termites, too, live in colonies numbering dozens to millions of individuals. As with ants, one or a few females do all the egg laying and a variety of worker types perform the main tasks necessary for colony maintenance. Both ants and termites use liquid food passed from one individual to another to regulate the type of worker a developing female becomes, and both communicate through the use of chemical signals, called pheromones; for example, pheromone trails are laid down to lead foragers to food and to recruit soldiers to battle.

Among the most amazing convergences between termites and ants (and also, in this case, some beetles) is the construction of underground fungus gardens. These insects invented agriculture tens of millions of years before we did! Although there are some differences between the farming practices of these different insects, the general plan is pretty much the same. Underground in a termite mound or ant nest, the insects bring in and plant fungus that is allowed to grow and then is harvested and eaten. The ant and termite workers carefully tend the garden, removing waste products, controlling pests, and eliminating other competing fungi species (they specialize on a particular fungal crop, treating the others as weeds). They even use antibiotics grown from bacteria housed in specialized regions on their body or in their guts to combat invading bacterial pests (ants use the same bacteria that we have used to produce the antibiotic streptomycin).

As this brief collection of examples suggests, convergent traits abound in the natural world. But it wasn't until 2003 that Conway Morris proposed that convergence is the dominant pattern in the biological world, rather than just a curiosity. His magnum opus, *Life's Solution: Inevitable Humans in a Lonely Universe*, offered 332 pages (plus 115 pages of endnotes) chockful of an extraordinary diversity of case studies of convergence from throughout life's expanse. Eight years later, George McGhee wrote a similar book, *Convergent Evolution: Limited Forms Most Beautiful*, slimmer than Conway Morris' book at 277 pages, but, if anything, even more jam-packed with examples. Even as I was drafting this chapter in 2015, a third tome appeared, Conway Morris' second offering, *The Runes of Evolution: How the Universe Became Self-Aware*, with another 303 pages of mostly new examples (and 158 pages of endnotes).

The net effect of these books is to overwhelm the reader with the breadth, depth, and sheer commonness of convergent evolution. It's everywhere! Think of just about any trait, and it's evolved multiple times, sometimes in distantly related organisms. Says Conway Morris, "Show me anything which has only evolved once and I'll . . . jump up and say 'no, I can give you another example.'"

For instance, McGhee notes that animals have evolved a variety of types of body armor to ward off predators. Turtles wear an impregnable fortress into which they retreat at times of duress, and functionally similar castles of bone evolved in a type of dinosaur (*Ankylosaurus*) and glyptodonts, a Volkswagen-sized extinct armadillo. Instead of a covering of bone, some animals enwrap themselves in sharp spines for defense. I've already mentioned the two independently derived types of porcupines; their approach is mirrored by echidnas (the only other egg-laying mammals beside the platypus, sometimes called "spiny anteaters"), hedgehogs, and hedgehog tenrecs from Madagascar, the latter two so similar in appearance that Richard Dawkins wondered why he even bothered to commission separate drawings for his book *Climbing Mount Improbable*.

Finally, although we think of armor as a physical defense to deter predators, noxious toxins in the skin can serve the same purpose. Such chemical defenses have evolved in nudibranchs (a type of marine mollusk akin to slugs); many types of beetles, butterflies, and other insects; pufferfish; frogs; salamanders; and a type of bird, the hooded pitohui, among many others.

Similarly, we mammals may be proud of our ability to give birth to live young (platypus and echidna excepted), but McGhee reports that live birth has evolved more than one hundred times in just lizards and snakes, not to mention repeatedly in fish, amphibians, sea stars, insects, and many other groups. Convergence extends even to the placenta—

the structure that transmits oxygen and nutrients from mother to embryo—which has evolved many times in both fish and lizards; in fact, the placenta of one lizard species is strikingly similar to that of some mammals.

And convergence isn't limited to the animal kingdom. To cite just one botanical example from McGhee's book, many plants rely on animals to transport their pollen from donor to recipient (pollen contains the plant equivalent of sperm). To do so, the plants need to attract their pollinators. In the case of hummingbirds, bright red apparently is irresistible. As a result, at least eighteen different types of hummingbird-pollinated plants have evolved bright red flowers.

Other plants, mostly in the Old World, have taken a different approach to procuring pollination services. Some species of flies and beetles lay their eggs in decomposing carcasses, and a number of plants—the corpse lily, the carrion flower, the Zulu giant carrion plant, among others—produce an odor that smells like rotting meat. The insects are

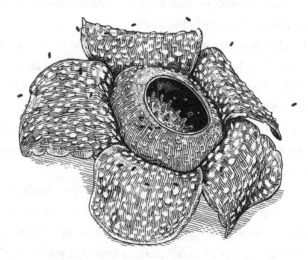

The corpse flower from Sumatra and Borneo is the largest flower in the world (yes, those are petals!). It attracts insects by emitting a smell similar to decaying flesh.

fooled and come poking around looking for a site to lay their eggs, picking up and depositing pollen as they do so. Seven different types of plants have evolved such odoriferous ways.

CONVERGENCE OF PARTICULAR TRAITS is fascinating, but most textbooks illustrate convergent evolution with examples of entire organisms that appear convergent. The iconic comparison is dolphins, sharks, and ichthyosaurs, all streamlined marine predators with flippers for forelimbs, a dorsal fin, a pointy snout, and a powerful propulsive tail capable of high-speed pursuit of their aquatic prey.

The other common textbook example comes from that upside-down land, Australia, where everything seems to be a little bit different. And at the top of the list are the mammals. I've already talked about Australia's penchant for evolutionary one-offs—the platypus, koala, and kangaroo leading the way. But there's another side of the coin. Much of the remainder of Australia's mammalian fauna is convergent with mammals elsewhere in the world.

After the dinosaurs perished, we mammals took over. In most of the world, it was the placenta-bearing mammals (the placentals) that grabbed the brass ring. Not in Australia, however. There it was the mammals that raise their young in external pouches—the marsupials— that reigned supreme. Despite this different evolutionary ancestry, the two mammal radiations produced many species that fill the same ecological niches in the same way.

Textbook writers like to line up Australian marsupials with their placental doppelgängers from elsewhere. Moles, flying squirrels, groundhogs—some of the parallels are so precise that if the marsupial form showed up in your North American backyard, you wouldn't think twice. I'm partial to the quoll, which not only looks and acts like a cat, but is said to make a good house pet. But perhaps the best example—

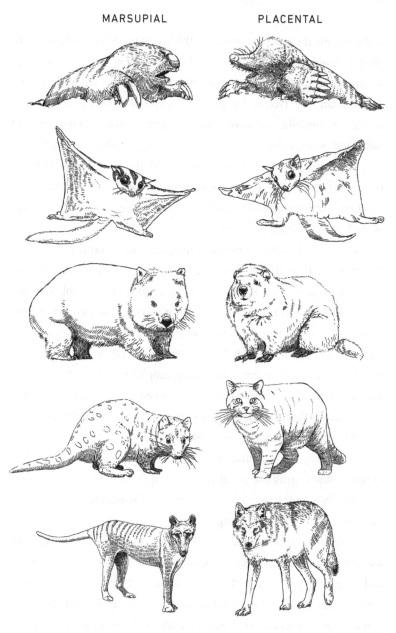

Australian marsupials and their convergent placental counterparts (from top to bottom): marsupial mole–mole; sugar glider–flying squirrel; wombat–groundhog; quoll–wild cat; thylacine–wolf

and certainly the most poignant—is the thylacine. A top carnivore with great resemblance to a wolf, I could easily see one of these creatures taking home Best in Show at Westminster, narrow snout and stiff tail notwithstanding. Decide for yourself on the doggishness of this species: go to YouTube and search for "thylacine"—you'll find a series of black-and-white videos of these animals wagging their tails, gnawing on a bone, jumping up and down, looking all the world like Buster, the family pet. Sadly, the thylacine is extinct, wiped out by Tasmanian ranchers a century ago—the eighty-year-old video footage shows some of the last individuals of the species.

Evolutionary copycats occur throughout the natural world. New World and Old World vultures are convergently ugly in their shared mortician's countenance. Australia's death adder is a member of the cobra family, but in appearance—and venom composition—it is a close match to the distantly related puff adder, a member of the viper family from Africa. Eel-like bodies have evolved not only in many types of fish, but also multiple times in aquatic amphibians and reptiles. Dry areas of Africa are covered with tough-skinned plants with sharp spines and no leaves, but they're euphorbs (members of the Euphorbiaceae), rather than the cacti of the New World.

This evolutionary emulation even crosses biological kingdoms. For example, tapeworms are members of the aptly named flatworm phylum that lives in the guts of vertebrates—including, maybe, you—and can grow up to thirty feet long, maybe even longer. At the front end, they have hooks and suckers that allow them to attach to the intestinal wall. In the neck region, they produce segments that contain embryos and have small projections thought to aid in nutrient absorption. New segments are produced toward the front of this region of the body, so that older segments are continually pushed to the rear end. Eventually, as the segment gets to the back end of the animal, the embryos are released or the entire segment breaks off into the intestinal void, and then you poop them out. If the tapeworm is lucky, you were doing your business in the great outdoors, and the embryos may find their way into their juvenile-phase host, an herbivore such as a grazing cow, in which they grow and develop. And if that cow is in turn eaten by a predator—such as you—without being cooked enough, then you've got a new intimate friend and the cycle begins again.

Although possibly spoiling your appetite, there's nothing particularly exceptional about this lifestyle—many other internal parasites live life similarly. What is unusual in this case is the story of dinoflagellates in the genus *Haplozoon*. Most dinoflagellates drift in the ocean, and many are photosynthetic (that is, they harness the Sun's energy to grow). But not *Haplozoon*. Despite being composed of only one cell, these organisms—which parasitize marine worms—have a body organization and life cycle paralleling that of tapeworms. For attachment to the intestinal wall, they have a sucker and hooks on the front end; for reproduction, they have egg-producing segments with small projections that originate mid-body and move posteriorly as new segments develop, eventually breaking off the end of the body, whence, just as with tapeworms, they are excreted out of the worm's body and set adrift to find their next host. What makes this case of convergence remarkable is that dinoflagellates last shared a common ancestor with tapeworms and other animals perhaps a billion years ago.

THE LIST OF EXAMPLES of convergent evolution is long and exotic, reaching into all corners of the biological world. But we really don't need to stray that far to see convergence—our own species provides many examples.

Homo sapiens emerged from Africa only 100,000 years ago, but in that short period we conquered the world, traveling and adapting to all four corners. And, in doing so, populations in different regions have occupied similar habitats—high on mountains in the Himalayas and the Andes, far to the north on several continents, in scorching deserts wherever they are found. The stage was set for convergence, and natural selection didn't disappoint.

The adaptive significance of variation in skin color among human populations has long been debated, but the field seems to be moving to a consensus that skin color reflects a balance between two factors. On the one hand, darker color, produced by a high melanin content in the skin, protects against ultraviolet radiation, which is particularly intense in equatorial regions. On the other hand, UV rays are important to the production of vitamin D. At high latitudes, where sunlight is less intense, lighter skin is favored to enhance the penetration of the vitamin-boosting UV rays.

Our species originated in Africa, which straddles the equator. As a result, the first humans likely were dark-skinned. This conclusion makes sense when viewed on a phylogeny; the earliest branches of the evolutionary tree—those that come off near the base—lead to the dark-skinned people of Africa. Later branching events lead to the light-skinned populations from Europe and Asia. These phylogenetic relationships leave little doubt that dark color is the ancestral condition in humans, from which lighter color evolved.

Geneticists have discovered the changes responsible for skin color and it turns out that the light coloration of people of Asian descent results from different mutations than those causing light color in Europeans. These genetic differences strongly suggest that light skin color evolved independently—convergently—in different populations as they colonized northern areas.* In turn, the ancestors of the aboriginal peo-

---

* The pattern of convergence extends to our nearest relatives, the Neanderthals, who also lived in northern areas and evolved light color by means of a mutation not found in any *Homo sapiens* population.

ple of Australia arrived Down Under about fifty thousand years ago, descended from presumably light-skinned Asians. Their dark coloration, thus, is convergent with the similar shade of African populations.

Another case of convergence among human populations involves the ability of adults to drink milk. One of the defining traits of mammals is the production of mother's milk to nourish their growing offspring. To digest it, young mammals produce an enzyme, lactase, which breaks down lactose, a sugar that is an important constituent of milk. Once a growing mammal is weaned, however, the gene that produces lactase shuts down because the enzyme is no longer needed. This occurs in most human populations, as well as in all other species of mammals. Cats, for example, are not adapted to drink milk, contrary to common wisdom. Feed an adult cat milk and it will have digestive upset, usually ending in diarrhea. The same is true for adults in most human populations—sixty-five percent of the adult human population is lactose-intolerant, and for them, drinking milk is an unpleasant experience.

The other third of the human population is more fortunate. How is it that those individuals, uniquely in the mammalian world, are able to continue drinking milk after weaning? Cows provide the answer.

Within the last few thousand years, human populations in disparate parts of the world—East Africa, the Middle East, northern Europe—began herding cattle. Why ranching occurred in those areas and not others is a subject of debate among anthropologists, but it is clear that these people took up cow-tending independently of each other.

With cows came a ready source of milk. To take advantage of this bounty, natural selection quickly found a solution, favoring genetic changes that kept the lactase gene turned on throughout life, rather than shutting off at an early age. Those of you who enjoy a cool glass of milk—as well as milk shakes, ice cream, and cottage cheese—can thank your cattle-herding ancestors for endowing you with the genetic machinery to do so. Although several human populations convergently evolved the same adaptive solution, genetic analysis reveals that they didn't do it in exactly the same way. Rather, different mutations—each with the same effect of keeping the lactase gene switched on—evolved in the different populations.

We humans aren't the only species in which multiple populations adapt in the same way. In fact, such within-species convergence is quite common: populations of the oldfield mouse have repeatedly evolved light-colored fur after colonizing dazzlingly white sand dunes; many populations of the Mexican tetra (a relative of the fish species familiar to aquarium keepers) have moved into underground caves and lost both their pigment and their eyes; many populations of the rough-skinned newt have evolved high levels of tetrodotoxin (the toxin found in blowfish and fugu) as a defense against their predator, the common garter snake; in turn, in many places garter snake populations have evolved physiological resistance to this toxin. I could go on and on. When closely related populations are exposed to the same selective environment, they tend to adapt in the same way.

SO FAR, I've talked about convergence between two species living in similar environments. This is an idea with deep historical roots. Darwin spoke of it in several places in *On the Origin of Species*, and evolutionary biologists have discussed it ever since. As I've detailed, the idea, though old, has blossomed in recent years as we've come to realize that convergence is much more common than we had appreciated.

Some related ideas, however, are more recent, with shallower tendrils burrowing back only a few decades. Darwin's idea focuses on a single selective factor and how multiple species evolve in the same way, but why should convergence be limited to one set of species adapting to the same environmental challenge? We know that in any given place, a wide variety of species exist, each adapted to its own ecological niche. If two places are very similar, might not natural selection produce an entire ensemble of convergent types, each adaptive form in one place paralleled by its convergent counterpart in the other? This is a much newer idea in evolutionary biology, one that has only been explored relatively recently. And much of this exploration has occurred on islands.

Replicated Reptiles

The author getting his start in herpetology

heck out the charming fellow in the photo. That's me, age thirteen, on a family trip to Miami to visit my great-aunt. And as I always did on those trips, I had gone outside to rummage through the luxuriant southern Florida foliage, looking for my favorite scaly beasts. On this occasion, my quest ended in success. The quarry: a little lizard, the Florida green anole, *Anolis carolinensis*.

Green anoles were (and still are) common in the pet trade, so when I wanted to conduct school science projects on lizards, they were the obvious subjects. In eighth grade, I investigated whether green anoles change color to match their background (contrary to common wisdom, they don't); in twelfth grade, I tried to figure out what cues trigger reproduction in the spring (the project failed, but the answer is increasing length of daylight).

So primed, I went off to college determined to study herpetology. When in my sophomore year a graduate student invited me to go to Jamaica to assist him in a field study of—what else?—*Anolis* lizards, it didn't require much thought (though I was disappointed when told that I needn't bring my tennis racket; apparently fieldwork was going to be not quite as I expected).

The green anole is North America's only native *Anolis* species,* but the genus is much more prolific elsewhere. More than sixty species occur on Cuba alone, and nearly 250 in mainland Central and South America. Jamaica, a tenth the size of Cuba, has seven species.

First stop was a marine laboratory on the island's north coast. Known as an excellent place for marine biologists to study Caribbean coral reefs, the lab had developed a second clientele thanks to its lush wooded grounds dripping with lizards. Forsaking flippers, masks, and snorkels, land-loving biologists were using the lab as a base for studying Jamaica's terrestrial fauna.

Once we arrived, it didn't take long to discover the lizards' diversity— they were everywhere, and not shy about drawing attention to themselves. Male anoles (and females in some species) possess a flap of skin under their throat called a dewlap. In repose, the dewlap is folded up, only visible as a ridge of skin running from near the tip of the jaw to the chest. But when the lizard has an announcement to make—"Get lost, buddy, this is my territory" or "Hey, ladies, come check me out. I'd make a good baby daddy"—out comes the dewlap, arcing downward from the jaw, forming a semicircle so large that the lizard often has to straighten its legs, pushing its body off the ground, to provide clearance.

---

* Ten others have been introduced from Caribbean islands in the last few decades.

The lizards occurred throughout the vegetation. The most common species was the most terrestrial, the Jamaican bush anole. These little fellows skittered back and forth from one bush to another or hung out low on a tree surveying their domain. Because they occurred near the ground, their drab brownness matched the background.

Up in the trees, however, was a different story. The most apparent was another species about the same size as the Jamaican bush. Found throughout the trees—high, low, on the trunk, scampering along branches—Graham's anole was as splendiferous as the bush anole was plain: sublime aquamarine washing over the head, torso, and forelegs, deepening to blue at the waist and cobalt on the tail; white dots and squiggly lines blanketing the body, transforming it into an organic QR code. Gorgeous already, males took it to the next level when they displayed their dewlaps, the bright orange contrasting with the varying shades of blue on their bodies.

Dazzling as Graham's anole is, it does not win the Jamaican lizard beauty pageant. That trophy goes to its larger tree-dwelling cousin, Garman's anole, known to Jamaicans as the green guana. The local appellation doesn't do justice to this verdant vision, cloaked in a tunic of chartreuse, a prehistoric dragon's spikey crest running down its back and a ring of yellow around its eye. Twice the length and eight times the weight of Graham's anole, the green guana is the Jamaican lizard king, a reigning despot as happy to dine on smaller relatives as on insects and fruit.

Yet another member of the anole clan lives in the trees, but this one has a different tailor. Rather than having the flashiest haberdashery, the Jamaican twig anole goes for understatement, so subtle in its grayish-white, brown-splotched suit that it blends right into its woody background, its slender form melding into the twigs on which it lies.

This scaly rainbow palette was a great introduction to the variety of Jamaican *Anolis* lizards, but it only told part of their evolutionary story. In fact, the species differ in many characteristics in addition to color.

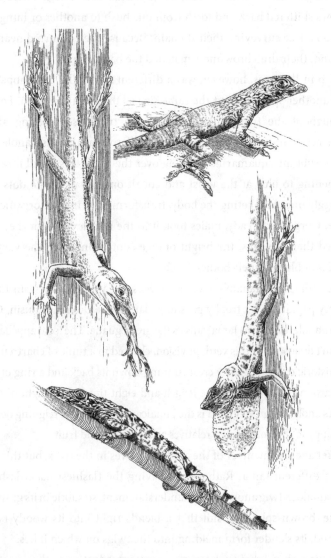

Jamaican habitat specialists (clockwise from top left): Garman's anole (green guana); Graham's anole; Jamaican bush anole; Jamaican twig anole

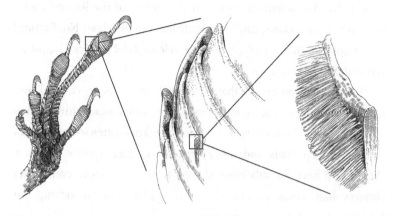

The toepads of an anole. Toepad scales are covered by millions of microscopic hairs.

Let's start with perhaps their most impressive attribute. On the tip of every toe, these lizards sport flat, oval-shaped pads. Flip the lizards onto their backs, and you'll notice that the undersides of the pads are composed of rectangular, slightly overlapping scales, arranged from small scales at the beginning of the pad, wider in the middle, and small ones again at the tip.

If you gently rub your finger along the pad, you'll feel a bit of resistance—even without moving, the pad is exerting a restraining force. Place the lizard on a horizontal plate of glass, its pads in contact with the surface. Then slowly start tilting up one end of the glass. As the glass approaches verticality, the lizard's body sags downward, but not the toepads, which remain firmly affixed, countering gravity's pull. Continue tilting, so that the glass passes the ninety-degree point, the lizard becoming ever more upside down, imitating Spider-Man, its four feet—or more precisely, its twenty toes—the only points of contact with the glass. Sometimes, even when the glass is completely inverted, the lizard retains its hold.

For decades, scientists pondered the source of the lizards' stickiness. Was it their claws, finding minute holes in the glass? No. Suction? Tiny hooks on the toepads, grappling Velcro-style? A sticky liquid secretion? None of the above.

It had long been known that the toepads of anoles are covered with millions of microscopic filaments called setae, each much thinner than a human hair. Another type of lizard, geckos—often seen at night scampering up walls and other vertical surfaces, mostly in warm areas—also have toepads covered with setae, and they're even better clingers than anoles. Presumably the setae play a role in sticking, but how they do so remained a mystery.

Finally, in 2000, a team of researchers figured it out, and the answer is something out of a science fiction story. Each of the millions of setae has free electrons on its surface, and under the right circumstances, these electrons can bond with electrons on the surface of glass or another object. These attachments—technically termed van der Waals forces—can be strong enough for a lizard to hang by a single toe or even catch itself as it's falling by touching its foot to a leaf. Try that, Spidey!

This, incidentally, has led to an entire new branch of engineering, sometimes called "gecko science." Think about the best adhesives we currently have, like superglue and duct tape. Either the stick can't be unstuck or it leaves a gooey mess. By contrast, not only can lizard toes undo their powerful adhesion, but when they remove themselves from a surface, they leave behind no residue. For this reason, a large and growing number of researchers have been trying to figure out how to translate the magic of the lizard's toes into products useful to us. The first, a new means of closing wounds, was introduced in 2008.

Now, back to the anoles of Jamaica. Like other anoles, they all have powerful, sticky toepads. But some are stickier than others. If you place a lizard on a gizmo called a force plate, and then gently pull the lizard backward, you can measure how much force the lizard's toepads are exerting against the force plate in its attempt to resist being pulled.

This method was pioneered by my former Ph.D. student Duncan Irschick, now a prominent researcher in his own right, and the results of his studies are clear: the amount of clinging force an anole can exert is proportional to the size of its toepads. And who's got the biggest toepads? Far and away, that would be the green guana, the largest of the Jamaican anoles. The other three species are about the same size, yet their toepads are quite different: the most arboreal of them, Graham's anole, has toepads nearly three times the size of the terrestrial bush anole.

There are two reasons more arboreal lizards need greater clinging ability. First, tree-inhabiting lizards more often use slippery, hard-to-cling-to surfaces, such as leaves or the very slick trunks of some tropical trees. Hanging on to such surfaces requires a stronger grip. Second, it's a much bigger deal to fall out of the top of a tree than from a perch a foot above the ground. Most anoles are small enough that they won't be harmed by such a fall, but there are very real costs to climbing back high into a tree, both in terms of energy expended and exposure to predators.

Comparing Jamaican anoles, a second aspect of their anatomy is also obviously different—the length of their legs. In this case, it is the terrestrial species, the bush anole, which has the upper hand, with hindlegs substantially longer than those of the other species (correcting for body size in the case of the green guana). At the other extreme, the twig anole, the dachshund of lizards, has very short legs and a long torso.

Among the Jamaican species and Caribbean anoles more generally, variation in leg length is related to where in the habitat a lizard lives. Species like the Jamaican bush anole that commonly use wide tree trunks and the ground have very long legs; those that use narrow twigs have much shorter appendages.

To find out why, we brought lizards into the lab and staged the Lizard Olympics. In the track-and-field competition, the first event was the two-meter dash in which lizards were encouraged to sprint up a narrow track, their speed calculated as they passed through regularly spaced infrared beams. Next up was the broad jump, in which the lizards were stimulated to jump by a gentle tap on the rump.

The results from these trials were straightforward: the longer a lizard's legs, the faster it could run and the farther it could jump. These findings made biomechanical sense, but left us unsatisfied—what was the advantage of short legs?

The pentathlon provided the answer. We again measured how fast lizards could run, but this time we sprinted them on five different surfaces ranging in diameter from very broad to very narrow. Our prediction was that the long-legged species would be zippiest on the broad surface, but the short-legged twig species would be quicker on the narrow rod, similar to the thin branches they use in nature.

Wrong! All the species ran more slowly on the narrow surface and the twig species was no faster than the others in that trial. Short legs clearly are not an adaptation for moving rapidly on narrow surfaces.

While we were running the lizards, we noticed that they sometimes stumbled or even completely fell off the surface. We recorded these miscues, but only later realized that this was the missing piece of the puzzle. On broad surfaces, none of the species had much trouble and mishaps occurred in only about twenty percent of the trials. But on the narrow surface, although the twig species continued to be unfazed, the long-legged species had tremendous difficulty, slipping or falling off completely in more than three-quarters of the trials. There was our answer: short legs are not advantageous for speed on narrow surfaces, but simply for being able to negotiate such substrates without difficulty.

In hindsight, this should have been our expectation. We spent hours observing lizards in their natural habitats. Longer-legged species frequently run at top speed to capture prey or elude attackers. By contrast, the twig species rely on stealth, not speed. Well camouflaged, they slowly creep up on prey; when they see a predator, they sidle around to the other side of a branch and melt away, step-by-step. For such a lifestyle, agility, rather than rapidity, is the key, and that's the advantage short legs confer.

THE ISLAND OF JAMAICA ORIGINALLY wasn't an island at all. Rather, it was part of Central America, attached to the Yucatán in Mexico or thereabouts. Approximately fifty million years ago, Jamaica and the continent went their separate ways, the newly born island drifting eastward into the Caribbean. While doing so, it sank under the waves, washing off its inhabitants, before reemerging millions of years later, a blank slate, poised for a new future. All that remained was for the first actors to find their way to the new evolutionary theater.

Subsequently, the ancestral Jamaican anole washed ashore, having drifted across the Caribbean from another island, probably Cuba. Over time, the initial species begat many descendant species,* and these species adapted to different parts of the habitat, evolving differences in color, toepads, limbs, and a variety of other features.

---

* The process of speciation—how an ancestral species diversifies into multiple descendant species—is one of the great questions in evolutionary biology and could be the subject of its own book (in fact, there have been several). Two populations are considered different species when they cannot, or will not, interbreed and produce fertile offspring. In the case of anoles, the dewlap seems to be an important part of the story—anoles can distinguish lizards of their own species from members of other species by the color and patterning of the dewlap, thus preventing hybridization. So, in anoles speciation is often the result of evolving different dewlap ornamentation. How that occurs, in anoles or in general, is a matter of debate. Geographic separation can be important because it allows the gene pools of two species to differentiate without constant intermingling. Natural selection pushing the populations in different directions can also hasten divergence. Evolutionary biologists are currently debating the relative importance of these two factors.

The result of this divergent evolution is a set of species, all descended from a geologically recent common ancestor, each adapted to its own ecological niche. Biologists have a term for this phenomenon—we call it "adaptive radiation"—and many think it is one of the most important aspects of evolutionary diversification.

Adaptive radiations are common and have long been studied by scientists. Indeed, Darwin's finches, so important in Darwin's thinking, are a textbook example. But adaptive radiation of anoles is special. To understand why, we'll have to go elsewhere in the Caribbean.

Five years after my trip, I was back in Jamaica, this time as a graduate student collecting data for my own dissertation. Again, I went to the marine lab and confounded a new crop of marine biologists by staying ashore and studying the lizards.

This time, however, Jamaica was only the first stop. After collecting the data I needed, I went island hopping. In the verdant Luquillo Mountain rainforest of Puerto Rico, I again found a variety of coexisting *Anolis* lizards, each species occupying its own part of the habitat. But not just any variety. Rather, the same habitat specialists that occur in Jamaica. Up in the canopy, a large species, similar in size and appearance to the green guana, and another, smaller green species moving around on the vegetation with very large toepads. Near the ground, a long-legged brown species with small toepads, adept at running fast and jumping far. And on the twigs, a well-camouflaged, short-legged species, incapable of rapid movement, but well designed for creeping along narrow surfaces.

If you didn't know better, you'd think these cross-island counterparts were close relatives, the twig anoles on the two islands evolutionary siblings, the green guanas kissing cousins, the long-legged ground runners the result of a recent evolutionary divergence. But they're not. The ecologically and anatomically diverse Jamaican species are all descended from a common Jamaican ancestor, more closely related to each other than to their ecological matches on Puerto Rico. Jamaica is one adaptive radiation, Puerto Rico another, and yet these two independent evolutionary unfoldings have produced a very similar set of habitat specialists (in fairness, I have to point out that the match isn't perfect—Puerto Rico has an additional type that hasn't evolved on Jamaica, a species adapted to living among grass blades).

The following year, I went to the Dominican Republic, one half of the island of Hispaniola (its island mate being the sad, but also lizard-rich, country of Haiti*). There, too, I encountered a variety of anole species, and once again, the now familiar doppelgängers, the same set of five habitat specialists that are found on Puerto Rico. And yet, once again, close ancestry is not the answer—appearances to the contrary, their DNA tells us that the twig anoles of the three islands are not closely related, nor are the brown ground species, nor any of the habitat specialists.

The twig anoles of Hispaniola (top), Puerto Rico (middle), and Cuba (bottom)

---

* I've never been to Haiti because years ago, when I thought about crossing the border from the Dominican Republic, I was advised that I probably wouldn't be killed, just thrown in jail. Some of my braver colleagues have gone there and seen marvelous things. Unfortunately, many of the most extraordinary Haitian species, lizards and others, are in dire jeopardy due to the wholesale destruction of Haiti's forests.

Finally, a few years later I was at last able to get permission to visit Cuba, no mean feat given the U.S. trade embargo and the Cuban suspicion that any gringo who proposed to work in the countryside was really a CIA agent plotting the invasion. I won't belabor the obvious: yet again the same set of habitat specialists, but independently evolved on Cuban soil.

The *Anolis* lizard communities of the four islands of the Greater Antilles are strikingly similar. Four types of habitat specialists are found on all four islands, in many cases the species of a particular type so similar across islands that they could be mistaken for the same species. An additional habitat specialist type is found on three of the islands, and yet two more types that I haven't mentioned occur only on Cuba and Hispaniola. Very few species occur on one island without a match having evolved on any of the others.

FOR A NUMBER OF YEARS, Greater Antillean anoles were the only well-documented case of what has come to be known as replicated adaptive radiation. Recently, however, a number of other examples have come to light, and it now appears that convergence of entire radiations is not as rare as once thought.

A recently uncovered example involves snails from some little-known Japanese islands. The snails are in the genus *Mandarina* and have no common name; their homes, collectively known as the Ogasawara Islands, are a set of thirty islands lying six hundred miles south of Tokyo. Remote, tiny (thirty-four square miles in total), and nearly uninhabited (population: 2,400), the islands' obscurity may seem well deserved. But what they lack in size and humanity, the Ogasawaras more than make up for in natural beauty. Recognized as a World Heritage Site in 2011, the islands are rich in biodiversity, so much so that they are sometimes referred to as the "Galápagos of the Orient."

And *Mandarina*, with nineteen species spread across the islands, is part of the reason for that reputation. Like anoles, coexisting species invariably differ in habitat use: some are found only in the trees, usually on leaves; others are semi-arboreal, occurring on both tree trunks and the ground; strictly terrestrial species divide into those that are found out in the open and those that live under shelter.

Concomitant with these habitat differences are differences in anatomy. Arboreal species are smaller than terrestrial ones, perhaps because it's not easy lugging a shell up a tree, especially when you don't have feet. The arboreal species also have a bigger shell opening, which allows a greater amount of the snail's body to come into contact with arboreal surfaces. The semi-arboreal species have particularly flat shells, which correspond with their penchant for wedging themselves into narrow nooks amidst the vegetation. The two ground types also differ, the more exposed one being more dully colored and flatter.

In other words, adaptive radiation, snail-style. And just like the Caribbean anoles, the snail radiation is replicated across the Ogasawaras. On several of the islands, the snails have diversified from a single ancestral species, producing a suite of different habitat specialists. And just as with the anoles, when we compare across the islands, we see that the same set of habitat specialists has evolved on each island; in some cases, the habitat counterparts on different islands are so convergent that they are indistinguishable in shell anatomy.

Another case of replicated adaptive radiation comes from the aerial realm. Of the more than five thousand species of mammals in the world today, one in five is a bat—1,240 furry, flying species at last count. A particularly successful group is the genus *Myotis*, the mouse-eared bats. There are more than one hundred species of mouse-ears; indeed, if you've seen a bat in North America, it may well have been the little brown bat, *Myotis lucifugus*, a very abundant and beneficial species that can gobble up to half its body weight in insects every day.

Murine ears notwithstanding, not all *Myotis* look alike. The traditional classification recognizes three main groups that differ both anatomically and in habits. Like the little brown, species in one group, subgenus *Selysius*, zip through the air in pursuit of aerial insect prey. They have small feet and a large membrane stretched between their hindlegs that they use like a catcher's mitt to snatch insects out of midair. Bats in the subgenus *Leuconoe* are narrow-winged and usually forage above water, sometimes even gaffing a fish from the surface with their long, hairy hindlegs. Lastly, species in the subgenus *Myotis* are large-bodied, big-eared, broad-winged bats that pluck prey off leaves, branches, or the ground.

It was accepted wisdom that these three subgenera represented different evolutionary groups, each type having evolved a single time and then given rise to many similar species that dispersed around the world. Then DNA comparisons turned the mouse-eared bat world upside down. By sequencing several genes from three-quarters of the world's *Myotis* species, researchers found that the traditional, anatomically based classification couldn't be more wrong. Just like *Anolis* lizards and *Mandarina* snails, anatomically and ecologically different bats living in the same region were more closely related to each other than they were to similar species in other regions. The story of a single evolution of each type followed by worldwide diaspora was disproven. Rather, mouse-eared bats seem to have repeatedly undergone adaptive radiation into the same three types in many parts of the world.

Stories like the anoles, *Mandarina* snails, and mouse-eared bats are becoming increasingly common. Apparently, environmental conditions are so similar from one place to another that natural selection drives the evolution of the same set of specialists in each place. These convergent pressures are so strong that the resulting species end up looking enough alike to be considered close relatives, rather than recognized as the product of independent evolutionary radiations.

I'll provide two last examples, the result of DNA-based studies on the inhabitants of Madagascar. A few years ago we learned that, contrary to what was once thought, the frogs of Madagascar are not closely related to their ecological counterparts in India. Rather, they have diversified in situ on the Red Island, producing burrowers, stream dwellers, and tree frogs very similar to those in India.

Madagascar's birds sing the same song. Despite great similarity to the African avifauna, DNA analysis recently revealed that the great diversity of perching birds in Madagascar is primarily the result of extensive within-island diversification. As we saw in Chapter One, the same story is true for the evolutionary radiation of Australian birds, which produced a panoply of disparate ecological specialists convergent with different species in the northern avifauna.

GOULD'S THOUGHT EXPERIMENT involves going back in time, starting evolution from the same initial conditions, and seeing if evolution unfolds in the same way. But there's another way to design this gedankenexperiment. Instead of replaying the tape in time, how about replaying it at the same time, but in different places? In other words: create identical environments in multiple locations and seed them with indistinguishable populations; then see if the populations follow the same evolutionary trajectories.

Of course, that's a gedankenexperiment, an ideal world in which anything is possible. In the real world, such an experiment is impossible because no two places outside the laboratory are truly identical, nor do they experience the same events through time.

Still, islands have been called the test tubes of evolution for a reason.

Their isolation gives them independence—what occurs on one island will not influence what happens on others (at least for species with poor dispersal abilities). And although never identical, many islands, particularly in the same general area, do tend to be fairly similar.

In the ideal world, the gedankenexperimenter would put identical lizard populations on four identical islands and observe how they evolve over millions of years. Is that utopian research program all that different from what really happened with *Anolis* lizards in the Greater Antilles? In the real world, this may be as close as nature has come to replaying Gould's tape.

And the results, of course, contradict Gould's thesis. The tape simultaneously played out on four Greater Antillean islands, and the lizard evolutionary outcome was very much the same on each. The fact that Jamaica is not the same as Cuba is not the same as Puerto Rico strengthens the rejection—the path of lizard evolution was not contingent upon the particular circumstances on each island.

Ditto for the land snails, mouse-eared bats, and many other organisms. Replicated adaptive radiations have occurred in many places, and more are regularly coming to light. Apparently evolution does repeat itself; independent rolls of the evolutionary dice can yield essentially the same outcomes, contingencies notwithstanding.

TAHITI, BERMUDA, MADEIRA, BALI. Everyone loves islands, but no one has nesiophilia—the inordinate fondness and hungering for islands— more than an evolutionary biologist. Darwin drew much of his inspiration from island stopovers on the fabled voyage of the *Beagle*; so, too, did Alfred Russel Wallace on his own perambulations through Southeast Asia. And ever since Darwin and Wallace jointly proposed their theory of evolution by natural selection, biologists have been returning to islands to gain fresh insights. Through their work, much of what we've learned about evolution over the course of the last century and a half has had its roots in island research.

What is it that keeps evolutionary biologists returning to islands? I can answer that in two ways. First, in the measured terms of a scientist: islands are replicated natural experiments of evolution. Each oceanic island or archipelago is a world unto itself, the evolutionary goings-on there independent of what has happened elsewhere. That means that by comparing one island to another, we can get a sense of evolutionary potential and predictability. Does evolution produce similar end results time and time again? For anoles and Ogasawara snails, it does, but how general is this result? How varied are the possible outcomes of evolutionary diversification? By comparing the similarities and differences among insular faunas and floras, island researchers aim to find out.

Here's a more fun answer: islands are cool! I'm not talking about the beautiful ambiance, but, rather, their extraordinary variety, the amazingly unusual plants and animals, so different from one place to the next. Take New Caledonia, for example, a mountainous corrugation of an island in the South Pacific, its coppery slopes covered with rainforests. In those jungles reside wonderful creatures: nocturnal geckos as big as my forearm; problem-solving crows that fashion their own tools out of palm fronds; *Amborella*, a quixotic plant from the dawn of flowers; a spiky-tailed giant land tortoise with horns on its head, and a land-living crocodile (the last two sadly wiped out by the first human New Caledonians).

Islands are full of oddities. But of all the crazy species that have evolved on islands, my favorite example of insular idiosyncrasy is straight out of Bedrock. Bedrock, Cobblestone County, that is, home of Fred Flintstone and Barney Rubble. Knowledgeable readers will recall the many technological advances of that visionary time—the foot-powered cars and pterodactyl-propelled airplanes, the bird's-beak record-player stylus and the construction crane *Brontosaurus*—but none was more biologically astute than the vacuum cleaner powered by the trunk of a pint-sized woolly mammoth on wheels.

No doubt, *The Flintstones'* Hollywood creators, the legendary team of William Hanna and Joseph Barbera, drew inspiration from the real (and oxymoronically named) dwarf mammoths that lived on the Greek island of Crete in the recent geological past. Less than four feet tall at the shoulder, these mini-mammoths weighed only about 500 pounds, perhaps three percent of their truly mammoth ancestors.

Or maybe Hanna and Barbera were thinking of another island, because Crete didn't have a monopoly on puny pachyderms. Indeed, tiny tuskers evolved on many islands around the world at many different times, some recent enough to have coexisted with modern humans: Malta, Corsica, St. Paul off the coast of Alaska; Flores, where they lived with Komodo dragons; even the Channel Islands off the coast of Southern California.

Three lessons can be drawn from this profusion of paltry proboscideans. First, there's a lot of money to be made in developing a line of petite jumbos. Who wouldn't want a Shetland pony–sized pachyderm

The pygmy elephant of Malta and Corsica, which survived until a few thousand years ago.

as a household pet? I'd take two! Second, this is yet another example of convergent evolution. Put an elephant on an island, wait a little while, and voilà, the incredible shrinking hulk.

But what is most important is that this is more than just another case of convergence, a particular type of animal or plant evolving in the same way in response to the same conditions. Rather, the evolution of small size in island elephants exemplifies a general evolutionary rule that extends beyond the trunked set and applies to many large mammals: dwarf hippopotami that evolved on several islands in the Mediterranean and elsewhere, deer that decreased in size by eighty-three percent after a few thousand years of insular living on the Isle of Jersey, even a three-and-a-half-foot-tall hominid (the "Hobbit," or, more technically, *Homo floresiensis*) that occurred on an Indonesian island as recently as seventeen thousand years ago. Put any large mammal on an island and it's likely to get small, sometimes downright tiny.

Another evolutionary trend on islands is for aerial animals to lose their flying ability—wings become smaller, some parts are lost, occasionally the entire wing disappears. Flightlessness in birds evolved numerous times on islands throughout the world: the dodo on Mauritius; rails (small birds that look like a cross between a chicken and a heron) on hundreds of islands; ibises on Hawaii and Reunion; parrots on multiple islands; the Galápagos cormorant; and in many others, including ducks, geese, parrots, owls, and a falcon. And the loss of flight is not limited to our feathered friends. Many island insects—particularly beetles, but others as well, including earwigs, moths, crickets, and wasps—have also been evolutionarily grounded.

Big mammals getting smaller, birds and insects losing their wings; one might think that islands are a land of evolutionary diminution. But let's not miss the forest for the trees. Some types of organisms increase in size on islands, and plants are a prime example. Trees generally do not produce seeds capable of surviving a long oceanic voyage, primarily because they're too big. As a result, few mainland trees make it to islands.* That means that newly formed islands generally lack high vegetation. There's a reason that trees are as tall as they are—skyscraping arboles are not shaded by others and thus can soak up the sunshine, maximizing their photosynthetic capabilities. Consequently, on an island lacking trees, any weed or flower slightly taller than the others would have an advantage; given enough time, one might expect normally small plants to evolve into tree-like forms. And that's exactly what happens. On island after island, species that on the mainland are shrubs, weeds, or small flowers evolve to become tall trees with bark and a central trunk, essentially indistinguishable from trees on the mainland.

Not all evolutionary regularities pertain to islands, however. Probably the two most famous examples of evolutionary predictability refer to how the size and shape of mammals and birds change with increasing distance from the equator. Think about bears. The two largest are Kodiak and polar bears, both denizens of the far north. Similarly, the largest cat is the Siberian tiger, substantially larger than other tiger subspecies and all other living felines. Bergmann's Rule, named for the nineteenth-century German biologist who described it, refers to the tendency for size to increase with latitude among individuals within a species or among closely related species.

Picture now the head of a polar bear and look at its ears. They're pretty small. So, too, are those of its fellow northern resident, the Arctic fox, whose legs are also noticeably short. In contrast, the familiar red fox from less frigid climes has bigger ears and longer legs; those of the fennec fox from African deserts are even more elongated. Again a

---

* Palms being an exception—the coconut is a seed that floats in the ocean and sprouts upon washing ashore.

trend, Allen's Rule, this one named for a nineteenth-century American biologist: the appendages of mammals and birds become shorter relative to body size at higher latitudes.

There's a lot of debate about these two rules, and they are clearly just generalizations with many exceptions. Still, the consensus is that temperature is the cause. In the north, endothermic animals—those animals that generate their own heat to maintain high body temperatures* —have to minimize loss of heat. Heat is produced in every cell of the animal's body and is lost through the animal's exterior. As animals get larger, their surface-to-volume ratio decreases, leading to less heat loss. Shorter appendages also limit heat loss. In contrast, in hot climates, the problem is the reverse—animals need to avoid overheating, so the larger surface-to-volume ratios of small body size and large appendages are more effective.

I'll mention one last general trend, first noted by Darwin a century and a half ago: domesticated animals tend to evolve a similar suite of traits. For example, many domesticated species commonly sport white patches of fur interspersed among areas with darker color. Such piebald patterning is seen in mice, rats, guinea pigs, rabbits, dogs, cats, foxes, mink, ferrets, pigs, reindeer, sheep, goats, cattle, horses, camels, alpaca, and guanacos. Floppy ears occur in breeds of rabbit, dog, fox, pig, sheep, goat, cattle, and donkey; curly tails have arisen in dogs, foxes, and pigs. Most domesticated species—including, I'm sorry to say, dogs and cats—have evolved smaller brains than their wild ancestors. And, of course, all domesticated species have become more docile.

Why these traits have repeatedly evolved is unknown. Except for

---

* Often referred to as "warm-blooded," but this term is not apt for several reasons, the most important being that even animals that do not produce much heat internally can maintain high body temperatures by basking in the Sun.

tameness, none of these characteristics has been the object of artificial selection programs. As a general rule, breeders have not tried to develop breeds with white splotches, nor have they intentionally favored flappy-eared goats or pigs with upward curving tails. Rather, these traits seem to evolve as a correlated consequence of selection on some other trait.

The evolution of these traits was illustrated by a long-running experiment in Siberia. In the late 1950s, the Russian geneticists Dmitry Belyaev and Lyudmila Trut purchased 130 silver foxes from an Estonian fur farm. They evaluated the foxes for their aggressiveness to humans and chose the most docile animals to breed. Belyaev and Trut applied the same criteria to their offspring and their offspring's offspring. Sixty years of such selection later, the resulting tail-wagging, belly-rub-loving, attention-seeking vulpines seem more dog than fox—in fact, they are now available for adoption (although shipping costs from Siberia are immense).

But alongside these behavioral changes, the foxes also evolved anatomical differences characteristic of the "domestication syndrome." Many now have a white patch on their forehead, when excited they roll their tails upward, and the ears of many puppies are as floppy as any Jack Russell terrier's.

Why these traits evolved in foxes, and in domesticated animals more generally, is not understood. The leading hypothesis is that selection for tame animals leads to hormonal changes that produce more docile behavior. But these hormones affect more than behavior; especially during pregnancy, the hormones play a role in regulating embryological development and thus have effects on anatomy. Consequently, the changes in hormones that result from selection on behavior have myriad other consequences, leading to the suite of traits that regularly evolve during the domestication process.

———

WIDESPREAD CONVERGENCE, replicated adaptive radiations, general evolutionary rules: the evidence for evolutionary determinism would seem overwhelming. Perhaps Conway Morris and company are correct, that evolution repeats itself in predictable ways.

But there is one problem. Much of the evidence—especially the long compendia of cases of convergence and replicated adaptive radiation—is gathered after the fact. Not only are these not experiments in which we test how predictable evolution is, they are not even an unbiased sample. These are the cases in which evolution did repeat itself. But in how many cases did it fail to do so?

chapter three

Evolutionary Idiosyncrasy

S nuffling through the underbrush, the shaggy little creature wanders through the sylvan night, sticking its nose in one place, then another, seeking the aroma of its soft-bodied dinner. The forest is dark and the pixie's eyesight poor, but long whiskers and a keen sense of smell allow it to get around. Threatened, it takes off at breakneck speed, barreling through the vegetation, ducking through holes, soon lost from sight.

An entirely unexceptional lifestyle. Many animals spend their nights cruising the forest floor, searching for small prey in a similar fashion: hedgehogs, shrews, weasels, to name a few, and bigger ones, too, like opossums and even pigs. The world is full of them.

But this one is different. All the others are hairy. This one's pelage is also soft, made up of millions of thin strands. But they're not hair. All the others move about on four legs and bear live young. Not this one.

Scratching, probing, sniffing, the animal often duets with its mate, calling back and forth, remaining in contact as they traverse their territory. And as the male calls, he identifies himself: "kee-wee, kee-wee."

We're in New Zealand, and this nocturnal insectivore is a bird, one with nubbins for wings, catlike whiskers, soft feathers, and, unlike any

Kiwi (illustrated by David Tuss)

other bird, nostrils on the tip of its beak. Many refer to it as an "honorary mammal."

The kiwi is not the only odd duck from New Zealand.* Most famous are giant flightless birds called moas, the largest towering nine feet above the ground and weighing 600 pounds. Others include a flightless parrot; a carnivorous parrot that attacks sheep; the adzebill, a stocky, flightless relative of coots with a massive, predatory beak; and the largest raptor ever, an eagle large enough to have preyed on moas. Birds aren't the only oddities. Other antipodean anomalies include tangled shrubs with branches on the outside and leaves tucked away in the interior; hamburger-sized snails; and an armored cricket as big as a rat, arguably the world's largest insect. New Zealand is chockful of unusual species.†

---

* There are actually five, closely related and very similar species of kiwis, all found in New Zealand.

† Or at least was. Unfortunately, many of the birds—including the moas, the adzebill, and the enormous eagle—went extinct in the last millennium, most as a result of human hunting and disruption of the ecosystem.

What is equally unusual, however, is what isn't there: mammals. There's scarcely a patch of fur on the islands. Not counting the seals that haul out on New Zealand's lovely beaches, the only native mammals are a trio of bat species, and even these are weird. Their hands transformed into wings, bats are generally clumsy on the ground. But not in New Zealand. The world's most terrestrial chiropterans, short-tailed bats scamper across the forest floor with great agility as they forage for insects, fruits, and nectar, prompting the noted biologist Jared Diamond to refer to them as "the bat family's attempt to produce a mouse."

Mammals have dominated the world's terrestrial ecosystems for the past fifty-five million years. New Zealand is a vision of an alternative world, one devoid of mammals. In their absence, birds have filled in, taking over the ecological roles normally played by mammals, but in unfamiliar ways. Perhaps if one squints, a kiwi could be equated with a shrew or a badger. But the dominant herbivores—the now-extinct moas and giant flightless geese—were a far cry from herds of antelope and deer, and a carnivorous parrot and an oversized, heavy-beaked coot are unusual substitutes for the familiar predatory complement of cats, wolves, bears, and weasels. Indeed, release from predation pressures is probably responsible for the supersizing of insects, snails, and other arthropods and the rodentization of bats. A bird-dominated evolutionary replay unfolds in very different ways than one ruled by mammals.

NEW ZEALAND is not alone in going its own way. Caribbean anoles and *Mandarina* snails notwithstanding, islands are rich in evolutionary eccentricity. Halfway around the globe, Cuba has its own peculiarities. The owl that was as tall as a first-grader and that may have eaten juvenile giant ground sloths is sadly gone (as are the sloths, one species as big as a gorilla), but the island is still home to a hummingbird as small as a bumblebee; the solenodon, an archaic mammal straight off the

Solenodon

pages of Dr. Seuss, with venomous saliva and a long, flexible, bewhis-kered schnoz; and beagle-sized guinea pig look-alikes that climb trees and produce copious, banana-shaped, green poops.

Even tiny islands have their unusual curiosities. Lord Howe Island, a five-and-a-half-square-mile crescent lying in the Tasman Sea, is home to six-inch-long black "tree lobsters" that, moniker notwithstanding, are bulky, oversized members of the usually wispy stick insect family. The Solomon Islands in the South Pacific harbor a lizard doing a mon-key imitation—the prehensile-tailed skink is a shiny, slender, two-and-a-half-foot-long lizard with a grasping tail that it uses to secure itself as it explores the forest canopy in search of fruit. Saint Helena is best known as the southern Atlantic Ocean island to which Napoleon was exiled. Less famously, until a few decades ago the island hosted a giant, three-and-a-half-inch-long earwig, a shiny black insect with an inch-long pair of forceps adorning its back end, vaguely reminiscent of a crea-ture from a *Star Trek* movie. And everyone's heard of the dodo on the Indian Ocean island of Mauritius, a flightless, fearless, fruit-gobbling pigeon the size of a tom turkey, three feet tall and weighing forty pounds.

Among small islands, however, the Hawaiian Islands take the prize for evolutionary oddballs: damselflies whose normally aquatic larvae live on land, voracious carnivorous caterpillars, fruit flies that have deserted their normal fruity fare for decaying plant matter, and other fruit flies that have hammer-shaped heads and defend their territories by head-butting as if they were bighorn sheep.

The Hawaiian plant world is equally off-kilter, headlined by the Ālula, which looks "like a bowling pin surmounted by a head of lettuce" (hence an alternative name, "cabbage-on-a-stick"). This three-foot-tall cliff-dwelling plant—"resembling no other plant in the world," according to one noted botanist—lives in crevices on the north faces of Kauai and

Hawaiian Ālula plant

Molokai, its bulbous bottom useful for swaying with the strong ocean winds, its tough succulent leaves an adaptation for the dry and salty conditions it endures.

And then there's Madagascar, sometimes called the eighth continent for the distinctiveness of its biota. We've already discussed the island's frogs and birds, but there's much more: a dwarf hippo; an adaptive radiation of lemurs, including a seventy-five-pounder that apparently hung upside down like a sloth and another that looked like a supersized koala;* ten-foot-tall, half-ton elephant birds (the heaviest birds ever to have lived); half of the world's species of chameleon, which propel their sticky tongues twice their body length to snare unsuspecting insect

---

* Both extirpated by the early human colonists of Madagascar in the last two millennia, along with all the other large lemurs.

prey; fossil frogs the size of an extra-large pizza; crocodiles that were vegetarians; a beetle with a giraffine neck. And the plants of Madagascar are no less unusual, including desert forests composed of tall, slender, spine-encrusted stalks and the stout baobab tree, which looks like it's been stuck into the ground upside down with roots coming out on top. And combining the animal and plant world, there's the orchid with a foot-long tube at the bottom of its flower and a corresponding moth with a proboscis equally elongated, several times the length of its body, just right for inserting into the tube and reaching the nectar at its base.*

Last, but certainly not least, are the wonders of Australia, the duck-billed platypus, kangaroo, and koala, unmatched by anything anywhere else in the world.

What do all these island oddities add up to? Islands reveal a glimpse of evolutionary alternative worlds, worlds that might have resulted if life had taken a different turn. What if mammals had been wiped out at the end of the Cretaceous along with the dinosaurs? New Zealand gives a suggestion of what might have been. Where would primate evolution have led if monkeys and apes hadn't evolved? Look no further than the diversity of lemurs, found nowhere else but on Madagascar.

Islands provide a grand cookbook of evolution. And the resulting concoctions inform us that there's no telling what will come out of the oven. Change the ingredients or the order in which they're added, turn up the heat, leave something out, use one pinch of salt instead of two, and the result may taste very different. Even when using the same recipe, seemingly innocuous events, like substituting one brand of flour for another or using your neighbor's kitchen instead of your own, may make a big difference. The island cookbook is replete with tales of contingency and chance, the diversity of outcomes suggesting that predict-

---

* The moth is famous in evolutionary biology circles because Darwin predicted its existence after studying the anatomy of the orchid.

ing what will evolve on any given island is very difficult. You just have to go there and look for yourself, and be ready to find almost anything.

Of course, evolutionary one-offs are not unique to islands. The natural world is full of extraordinary plants and animals that have no evolutionary parallels. Consider the elephant: what other animal uses its nose to pick up objects, give itself a dust bath, and lovingly caress a family member? Or the archerfish, whose visual system and oral anatomy allow it to shoot a precisely aimed jet of water, capable of knocking insect prey off branches and into the water. Not to be outdone in long-range attack, bolas spiders produce a long silk thread with a sticky blob on the end, which the spider swings at prey like a gaucho, the globule adhering to any unfortunate moth that it hits. And moving from hunting to reproductive tactics, there's the male anglerfish, vastly smaller than the female, who bites the female and secretes an enzyme that digests his lips and her skin, fusing their bodies together; ultimately, most of the rest of the male's anatomy disappears as well, leaving only the testes with a job to fulfill. And, of course, big-brained, tool-using bipeds. The biosphere is replete with species uniquely adapted to their mode of life.

CONWAY MORRIS and his colleagues have made long lists of examples of convergence, but it would be just as easy to make comparable catalogs of species without counterparts. We can easily understand convergence, species adapting in the same way to similar circumstances. But what's so special about the evolutionary one-offs? Why haven't other species convergently evolved similar adaptations?

One possibility is that these species occur in unique environments. Perhaps they have no analogs because no other species has experienced a similar environment. This, possibly, explains the koala. Its entire lifestyle revolves around living in eucalyptus trees and eating their leaves, which are loaded with poisonous compounds. As a result, the koala's digestive system is extremely long, providing ample time to slowly detoxify the leaves and extract the nutrients. This slow passage, combined with the low nutrient value of the leaves, means that koalas are on a tight budget, and as a result they minimize energy expenditures, sleeping away most of the day. Eucalyptus trees naturally occur only in Australia, so maybe the singularity of the koala reflects the uniqueness of its environment.

But I suspect this is not the explanation in most cases. Platypuses* occur in streams and ponds in eastern Australia, where they eat crayfish and other aquatic invertebrates that they locate by rooting around on the bottom, sensing their prey with electroreceptors located on their bills. When they're not out paddling around, they retire to their rest chambers at the end of long burrows dug into the stream bank.

The platypus lifestyle would seem to be possible in many places beside Australia. The streams they occupy are much like the creek that ran behind my friend's house when I was growing up in Saint Louis. Certainly, North America is full of crayfish-packed streams, many in areas with climates similar to the one the platypus experiences, and seemingly with no worse predators than in Australian waterways. So where's our platypus doppelgänger? Why hasn't anything like the platypus evolved anywhere else? Or the kangaroo, or any of the other examples I listed, all of which occupy habitats that occur elsewhere?

The other explanation for evolutionary one-offs is that natural selection is either not as predictable or as powerful as some make it out to be. That is, even when species experience identical environments, they might not evolve in the same way.

---

* There is controversy over the proper plural form for the platypus, with the contenders being "platypus" and "platypuses." Two other possibilities, however, are not in contention. Platypus is derived from Greek ("flat-footed"), so the Latin plural "platypi" is incorrect. In theory, the Greek plural "platypodes" would be correct, but this usage has never been adopted.

A key reason for lack of convergence is that there may be more than one way to adapt to a problem posed by the environment. Think about the way vertebrate animals swim. Many use their tail for thrust, but not all tails are the same. Fish tails are vertically flattened and are moved back and forth. Crocodiles swim in the same way. But whale tails are horizontally flattened and are moved up and down. Other animals, like eels and sea snakes, undulate their entire bodies. A few birds, such as cormorants and loons, can move speedily underwater by paddling ferociously with their web-footed hindlimbs. On the other hand, some species swim using modified forelimbs, like the flippers of sea lions and the wings of penguins. However, the most surprising swimmer may be the tree sloth, whose long forelimbs, evolved as an adaptation for hanging upside down, can produce a passing imitation of the Australian crawl. Invertebrates offer even more means of rapid aquatic locomotion, such as the jet propulsion of octopuses and squid.

This list of different ways to move quickly through water brings up the obvious question: to be considered convergent, how similar do the traits of two species have to be? Squid and dolphins use very different anatomical structures to move rapidly through the water—there's no question that they are not convergent. The foot-propelled locomotion of some aquatic birds is yet another non-convergent means of rapid underwater propulsion.

Other examples, however, are not so clear-cut. What about the tail flukes of cetaceans and sharks, similar in design and operation, but one horizontal and moving up and down, the other vertical and swept left and right? Do these features represent slight variations on a convergent theme, or non-convergent solutions producing the same functional outcome? I suspect that most people would consider horizontal and vertical tail flukes to be fundamentally the same solution.

Let's move back a step, to a trait that produces the same functional result, but exhibits greater anatomical variation among species. Pow-

ered flight evolved three times in vertebrates: in bats, birds, and pterosaurs (the large reptiles that conquered the sky during the Age of
Dinosaurs). All three modified their forearms into wings and fly—or
flew in the case of pterosaurs—in fundamentally the same way, by flap-
ping a lightweight structure downward to produce lift and forward
thrust.

But closer examination reveals that the wings of these flying vertebrates are built in very different ways. The most obvious difference is
the aerodynamic surface itself. Birds use feathers, individually growing from the arm bones. In contrast, the airfoil of bats and pterosaurs
consists of thin, but very strong skin stretched between finger bones
and the body, in some cases even attaching to the hindlegs. The skeletal anatomy of the wings of these three groups of fliers is also very
different.

*Bats (top), birds (middle), and pterosaurs (bottom) evolved wings by elongating
different elements of the forelimb. In addition, the wing surface of bats and
pterosaurs is composed of skin, whereas birds use feathers.*

So, are the forearm-modified wings of birds, bats, and pterosaurs convergent adaptations for powered flight that are built in different ways? Or do they represent alternative, non-convergent means of evolving powered flight?

One more example. The largest fish in the sea is the sixty-foot-plus whale shark, so named because it looks a lot like the baleen whales. Like the great whales, it is a filter feeder, gulping enormous quantities of water into its massive mouth and filtering out the minute food upon which it feeds. But that's where the resemblance ends. The baleen whales—blue, humpback, gray, and others—strain their prey by pushing water through stiff plates of comb-like baleen that form a curtain hanging from their upper jaws. Any food particle larger than the tiny gaps in the baleen is trapped on the inner surface of the baleen curtain and then ingested. By contrast, whale sharks filter their food in a very different way. Water is pushed out through gill slits positioned on either side at the back of the head. Filter pads made of cartilage are positioned in the gill slits in such a way that the water rushes between the pads, through the gills, and out into the ocean, but food particles continue moving backward past the gill slits, forming a mass in the throat that is subsequently swallowed. So, baleen whales and whale sharks are both large aquatic creatures that use enormous mouths to take in water and filter out small prey. Yet, the precise structure that does the filtering is built, placed, and functions differently. Are these convergent or non-convergent adaptations for filter feeding?

Where one draws the line between convergence and non-convergence among structures that are grossly similar and produce the same functional advantage is arbitrary. My inclination is to consider the wings of birds, bats, and pterosaurs to be convergent. Similarly, I view baleen whales and whale sharks convergent overall because both are large-mouthed, filter-feeding planktivores; however, I consider their filter-feeding structures to be non-convergent, alternative adaptations for filter feeding. But, really, there is no right or wrong answer in cases like these.

In other cases, though, species can adapt by evolving clearly different, non-convergent phenotypes that produce the same functional capabilities. My favorite example of this phenomenon concerns the subterranean lifestyle of rodents. More than 250 species in the rat clan spend much of their lives underground, moving through self-constructed tunnels. Such burrowing behavior has evolved repeatedly in the Rodentia, but it has been accomplished in different ways. Many rodents dig in the standard way, using their forelimbs to loosen dirt and throw it behind them. The forelimbs of such species are stout and highly muscular; the claws long and strong. Other species use their teeth rather than their claws for soil removal. As you might expect, the teeth of these species are long and protruding, even by rodential standards, and the jaw muscles and skulls are massively constructed. Most dentition diggers get rid of the soil by kicking it backward with their forelimbs, but yet another variation occurs in some rodent species, which pack the loosened soil into the tunnel wall with upward thrusts of their elongated, spade-like snouts. The diverse anatomies of these diggers are a clear illustration of non-convergent adaptations that produce the same functional outcome.

Non-convergence can result for another reason. Often there are different functional ways to adapt to an environmental condition. As an example, consider how potential prey species may adapt to the presence of a predator such as lions. One option is to evolve great sprinting ability to outrun them, but there are other options, too, like camouflage, passive defense, or active defense. The resulting adaptations are decidedly non-convergent, encompassing the horns of the cape buffalo, the body armor of the pangolin and tortoise, the long legs of the impala, the spines of the porcupine, the venom and precision projection of the spitting cobra, and the dappled pelage of the bushbuck.

Multiple solutions to the same selective problem are not limited to defense. Cheetahs and African wild dogs hunt the same prey, but the cat does so by short bursts of great speed, whereas wild dogs run more slowly, but for long periods, exhausting their prey and eventually bringing them down. The adaptations of the two are correspondingly different: the extremely long legs and flexible spine of the cheetah allow it to attain speeds of seventy miles per hour; the great stamina of wild dogs allows them to maintain a steady pace of thirty miles per hour for long enough to fatigue their prey (cheetahs can only sustain their sprints for a short distance).

Or consider the adaptations animals have to obtain nectar. Plants produce the often sweet-smelling, sugary liquid to bribe insects, birds, and other animals to aid in their reproductive process. When an animal sticks its head or entire body into the flower to lap up the nectar, it gets covered with pollen. When the animal goes to the next flower, some of the pollen falls off, leading to the fertilization of the plant's ovules.

Many flowers have very long tubes with the nectar at the bottom— in this way, the plant can limit who gets the pollen to one or a few particular species that are well adapted to using that plant, such as moths with their long proboscises and hummingbirds with similarly long beaks and tongues. Such species, because of their adaptations, probably don't visit many other types of flowers, limiting the extent that the pollen will fall off in a plant of a different species and thus be wasted.

But not all nectarivores play by the rules. Some species of insects, birds, and mammals chew a hole in the base of the flower, bypassing the petals and their pollen and consequently not upholding their end of the coevolutionary bargain. To do so, these nectar thieves possess very different adaptations. Rather than long tongues and mouthparts needed to get to the bottom of long tubes, these species evolve features that enhance their ability to tear through the flower wall. Some humming-birds have serrated edges on their bills for this purpose; the bird aptly named the flowerpiercer has a sharp hook on the tip of its upper bill used to slice through flowers.

What we see in these many examples is that there are often multiple evolutionary options to respond to a challenge posed by the environment. But just because there are multiple possibilities doesn't mean that all, or even more than one, will evolve. Conway Morris and crew argue that usually one option is superior to the others, and that is why the same trait evolves convergently, time and time again. Yet, convergence doesn't always occur. Why wouldn't natural selection favor the same trait every time?

It may be that two (or more) traits are equivalent. Being camou-flaged or fleeing at top speed may be equally successful means of elud-ing predators. Or maybe one approach is more successful than another for a particular purpose, but with other costs that counterbalance its advantage. Rapidly fleeing from an approaching predator may be a bet-ter means of escape, but being camouflaged may enhance the ability of an animal like a snake to ambush its own prey. When survival and re-production are totaled, individuals that are camouflaged may be just as successful as those that rely on speed in reproducing and passing their genes on to the next generation. As a result, natural selection would not necessarily favor one over the other. Which trait evolves might be a matter of chance, a function of which mutation occurs first in the pop-ulation once it is subject to predation.

Alternatively, which trait evolves might be contingent on the initial phenotype and genotype of the species. A species that was generally active might be predisposed to evolve whatever traits produce greater speed when faced with a new predator, whereas a more sedentary spe-cies might instead evolve camouflage. Neither option is superior to the other, but the evolutionary outcome might strongly depend on the initial conditions.

It may also be that one solution actually is superior, but in some cases it's easier to evolve a suboptimal solution. The French scientist François Jacob, who received the Nobel Prize for his research on how DNA works, proposed an analogy to explain why natural selection would not always lead to the evolution of a perfectly designed organism. Natural selection, Jacob said, is not like an engineer, constructing the optimal solution to the problem at hand. Rather, he said, think about a tinkerer, a handyman who makes use of whatever materials are available to fashion whatever solution is feasible—not the best solution possible, but the best attainable under the circumstances.

Now think of a bird species that finds itself in an area with a lake full of slow fish. It may start diving into the water for a piscine meal and in time may begin to adapt to a more aquatic existence, evolving extra-large and powerful hindfeet like a cormorant or sculpting its wings into flippers à la penguins. Let's suppose that the best way to swim quickly and with agility is to power through the water with a strong, muscular tail, pushing back and forth or up and down—that's what the fastest swimmers do. But birds don't have long tails—they lost them early in their evolutionary history, more than one hundred million years ago, leaving only a tiny residuum of fused bones (the "tails" of birds are composed only of feathers, not bone). I'm not saying that re-evolving a long tail is impossible, but natural selection, the tinkerer, probably wouldn't take that route. The bird already has wings and feet that can provide some propulsive force. It seems much more likely that natural selection would work to enhance the swimming performance of these pre-existing structures than to evolve a new structure completely from scratch, even if ultimately a remodeled bird with a bony tail—looking something like a cross between a loon and a crocodile— might have been a better swimmer.

But, still, if a bird-crocodile would be better adapted—a superior, faster swimmer—why wouldn't the swimming bird continue to evolve in that direction? The answer may be that sometimes you can't get there from here: evolving from one adaptive form to another may be difficult because intermediate conditions are inferior. A long, powerful tail may be great for rapid propulsion, but a short flap of a tail may just get in the way, actually decreasing swimming performance. Natural selection has no foresight—it won't favor a detrimental feature just because it is an early step on a path leading to an ultimately superior condition. Rather, for a feature to evolve by natural selection, every step along the way must be an improvement on what came before it—natural selection will never favor a worse condition, even if it's only a transient evolutionary phase.

As a consequence, species may end up stuck with suboptimal adaptations. For whatever reason, their ancestors didn't embark on the best road to adaptation. Natural selection pushed the species along, and it ended up adapted, but not as well as it might have been. This reasoning emphasizes the role that contingency may play in determining evolutionary direction and why, as a result, species may fail to converge when faced with identical environmental conditions. Differences among species in their ancestral genotype and phenotype or which mutation happens to occur first may lead species to adapt in different ways, even sometimes ending with inferior adaptations.

BY THE SAME LOGIC, we might expect that the more similar two ancestral species are, the more likely they would be to evolve in the same way when facing similar selective conditions. And that is exactly what happens. It's no coincidence that the best examples of repeated convergence are among closely related species. *Anolis* lizards have evolved the same suite of habitat specialists four times, yet no other type of island lizard has converged with anoles. The two nearly identical species of beaked sea snakes are in the same genus. Sticky toepads have evolved eleven times in geckos, and only two other times among the more than six thousand other species of lizards. Not all cases of convergence involve near evolutionary kin, but a recent statistical analysis confirmed that convergence is more common among closely related species.

The effect of close relatedness is particularly obvious when comparing populations of the same species, which often repeatedly evolve the same trait when exposed to similar environmental conditions. I provided many examples in Chapter One—mice on sand dunes, cavefish, toxic newts and their garter snake predators, humans—and will add just one more here.

The three-spined stickleback is a small fish, usually about two inches in length, found in coastal waters in northern parts of most of the Northern Hemisphere. The most prominent features of this slender fish, eponymously enough, are the three tall spines arranged in a line along its back in front of the dorsal fin, matched by another spine underneath where the pelvic fin should be. Predation must be a great threat to these ocean dwellers, because not only can these spines be locked into an erect position, but the fishes' sides are armored in bony plates, as many as forty in some individuals.

Much of the Northern Hemisphere was buried under glaciers during the last Ice Age. When the ice melted about ten thousand years ago, new streams emptied into the ocean. Like salmon, sticklebacks reproduce in freshwater, and local populations quickly took advantage of these new breeding grounds.

But then the landscape changed again. When a pile of ice a mile high is stacked on the ground, the land sags under the weight. But once the ice is removed, the land slowly rebounds, growing in elevation. And as that happened in what is now Canada, some streams were cut off from the ocean and converted into lakes. And trapped therein were formerly marine sticklebacks.

This happened to countless thousands of rivers, brooks, and rills, particularly along the western coast of North America. These waterways were geologically new and not well populated—few other oceanic fish species accompanied the sticklebacks upstream. The result is that these new lake populations found themselves in a novel environment, one mostly lacking predatory fish.

Consequently, the lake stickleback populations, each isolated in its own tub and evolving independently, changed in parallel. Why waste energy and resources building defenses against non-existent fish predators? The populations convergently lost most of their body armor and their spines shrank. Genetic studies showed that this evolutionary parallelism extended to the genome; across lake populations, the same genetic changes were responsible for evolutionary change in armor and spines.

The prevalence of convergence among closely related populations and species is easy to understand. Close relatives tend to be similar genetically, so selection is likely to have the same genetic systems on which to work. Moreover, relatives tend to be similar in many phenotypic attributes.

Because of these similarities, closely related species and populations share the same evolutionary predispositions, more likely to evolve in some ways than in others. Some evolutionary biologists refer to these predispositions as constraints or evolutionary biases. These biases could operate in a number of ways. The most obvious is the genetic similarity of close relatives, presenting the same target to natural selection, but more subtle biases could occur as well. An evolved trait in an ancestor could preclude some evolutionary options, forcing evolution to occur in a limited number of other ways among the species' descendants. Alternatively, an ancestor could evolve a trait that paves the way for the evolution of a second trait. Such potentiation, as it is now referred to by molecular biologists, would have the effect that closely related species would all evolve the second trait, one unlikely to arise in species not descended from that ancestor.

For all these reasons, related species are more likely to convergently evolve the same traits when faced with similar selective pressures. That's not to say that distant relatives can't converge—it certainly does happen, just less frequently.

THIS IS A GOOD PLACE to digress briefly and point out that convergence need not reflect adaptation to the same circumstances or even be a result of adaptation at all. The reason is that natural selection is not the only process that causes traits to evolve. Occasionally, traits evolve randomly, particularly in small populations. A trait may also evolve because it is linked genetically to another trait that is favored by selection or as a result of persistent immigration from another population. Consequently, convergent evolution could occur coincidentally if two populations happen to evolve the same trait for non-adaptive reasons. Such non-adaptive convergence may be most prevalent among related populations or species because of their shared evolutionary predispositions.

Salamander toes provide an example. Many salamanders have convergently evolved to have four digits instead of the ancestral complement of five. An adult salamander's toe number is determined by how many cells occur in the developing limb early in embryological development. Anything that reduces the number of cells in the limb bud—such as an increase in cell size or an overall reduction in body size—can lead to a reduction in the number of toes. We have no evidence that this convergent evolutionary reduction has been driven by natural selection: four-toed species don't occur in particular habitats and there is no benefit to having fewer toes (as far as we're aware). The more likely explanation is that toe reduction has occurred convergently for non-adaptive reasons, perhaps some species randomly evolving larger cell size and others being selected for small body size.

Ideally, we would directly test the hypothesis that natural selection has guided convergence. Relevant data can come from direct measurements of natural selection, detailed analysis of what benefits, if any, a trait confers, and knowledge of the species' evolutionary history. Even just the observation that a trait has evolved repeatedly in the same environmental circumstances suggests an adaptive explanation—a correlation between trait evolution and environment wouldn't be expected without the involvement of natural selection. Unfortunately, sometimes we don't have any relevant information.

Consider *T. rex*. Fearsome and terrible as it was, the tyrant king had one shortcoming: its arms. Puny and two-fingered, its forelimbs couldn't even reach its mouth. Scientists have put forward all manner of explanation, one crazier than the next. Maybe the super-predator fed in such a frenzy that its arms evolved to be short so it wouldn't accidentally bite them off and eat them. Perhaps the little limbs were used for pushing off the ground to get up after a nap. Possibly, male *T. rex* needed shorter arms to better titillate their mates. Needless to say, none of these ideas has gained support.

Recently, paleontologists discovered a new species of theropod dinosaur, *Gualicho shinyae*, that sported similarly feeble, double-digited appendages. Even though we don't understand why this trait evolved in either species, one of the authors of the paper said, "obviously there was some adaptive advantage because we see it multiple times in different lineages of theropods."

But maybe it's not so obvious. Convergent evolution doesn't necessarily prove that a shared trait is the result of natural selection. Maybe *T. rex* and *G. shinyae* both just happened to evolve diminutive forelimbs by chance. If we knew why small limbs with two digits evolved, what advantage they provided, or why natural selection favored them, we would have reason to think the convergence was adaptive. But absent any data, we can't just assume that natural selection is the cause.

I WANT TO CONCLUDE with one last grubby case study that illustrates several of the different routes to evolutionary non-convergence. This example concerns the species that eat insect larvae found in wood. Everyone knows the rat-a-tat-tat as a woodpecker jackhammers a hole into a tree, launching its head at high speed against the tree as many as twenty times per second.* What many don't know, however, is how Woody gets the grub out once he's gotten to it. The way he does so is by inserting his long, bristly tongue—so lengthy that when not being used, it's wrapped around the back of his braincase—deep into the hole, using the prickles to snag the prey and pull it out.

A pretty nifty trick, for sure, but it turns out that members of the Picidae don't have a monopoly on grub excavation. Woodpeckers have a nearly worldwide distribution, but they are not adept at dispersing across the ocean, and hence are absent from Australia and many islands. In their absence, other species have evolved to take over the grub-feasting niche, but none have done so in the woodpeckerian way. Rather, what we see instead of convergence is a set of different solutions to the problem of extracting insect prey from wood.

In the Hawaiian islands, the grubster is a lovely bird, yellow-headed in males and olive in females, with an extraordinary bill. Or bills, I should say, as the upper and lower parts are quite different. The lower bill is short, stout, and straight and used to excavate a hole, woodpecker-style. But instead of a tongue for extraction, the 'Akiapola'au (pro-

---

* And for that reason, now a subject of study by concussion researchers.

Different ways of adapting to eating grubs (from top): huia, `Akiapola`au, woodpecker, and woodpecker finch.

nounced ah-kee-ah-POH-LAH-OW) has a slender and deeply down-curved upper bill, twice the length of the lower bill, which can reach deep into a hole and pry out its larval prey.

In New Zealand, such versatility is apparently too much for any one bird, so the huia (HOO-ya) took a different tack, dividing the tasks between the sexes. The male was the more woodpeckerish of the couple, sporting a robust beak that he used to chisel into rotting wood for grubs; in contrast, both halves of the female's bill copy the 'Akiapola'au's top half, the thin, deeply curved bills used for extracting prey from deep crevices. At one time, it was thought that couples acted as a team, the male drilling, the female snagging, but this idea appears to have been a misreading of the original scientific report; now it's thought that the two sexes foraged separately. Alas, the species became extinct sometime in the last century, so more detailed study is no longer possible.

Perhaps the most remarkable avian adaptation to this way of life is shown by the bird with the least remarkable bill. The woodpecker finch of the Galápagos, a member of the Darwin's finch tribe, has a pretty standard, uncurved bill, neither particularly stout nor slender, long nor short. Not tough enough to hammer nor delicate enough to probe. That doesn't matter, however, because the woodpecker finch doesn't use its bill to extract prey. At least not directly. Rather, like a chimpanzee fishing termites out of their mound, the finch holds a stick of just the right size in its bill, poling it into a hole or crevice, wiggling and finagling, probing and prodding, until at last the larva is coaxed out and quickly consumed. Also like chimps (and like New Caledonian crows and, of course, us), these finches do not blindly use whatever stick is lying around. Rather, they carefully choose an implement, sometimes precisely trimming and tailoring it to fashion a tool just right for the job.

From these examples, you might think that, as a way of life, digging

Aye-aye

grubs out of trees is for the birds. But you'd be wrong. On Madagascar, that island of evolutionary marvels, the grub-extracting species may be the most extraordinary. There, the woodpecker niche is occupied not by a bird, but by a primate. And what a primate it is! The size of a house cat, the nocturnal aye-aye looks like something straight out of a horror movie. Glowing yellow eyes; big, floppy, black leathery ears set against a light-colored face; enormous forehead with a narrow, short muzzle; thin gray hair growing wildly from the top and side of its head—the animal looks like a sinister cross between Albert Einstein and Yoda. Unlike either the physicist or the grand master of the Jedi, however, aye-ayes have a pair of enormous, ever-growing incisor teeth and—the trait of which nightmares are made, no doubt responsible for many of the ancient Malagasy beliefs about the creatures' magical powers—an elongate, skeletal middle finger capable of rotating in any direction.

How the aye-aye gets its grub is truly amazing. It starts by tapping its long finger against the tree trunk, large ears functioning like radar dishes to interpret the returning sound, listening for the telltale percussion of an empty internal space in the wood. Once the potential grub tunnel is located, the aye-aye then deploys its forward-angled incisors, channeling straight through the wood to the void. After the cavity is opened, in goes the long finger, twisting first one way, then the other, until the grub is hooked by the claw and extracted. Who needs a fancy bill when a long finger and teeth will do the job?

Why has natural selection produced different solutions to the same grubby problem? It's conceivable that insect larvae differ from one place to the next, that the best way to catch continental grubs is the woodpecker way, whereas Galápagos grubs are best finagled by twigs and Malagasy larvae are particularly vulnerable to detection by big-eared primates. We can't rule out this Panglossian possibility that each species has evolved the uniquely optimal way of catching local grubs, but two alternative explanations are more likely.

One scenario is that the differences evolved by random chance. Perhaps the ancestor of woodpeckers gained a mutation that led to a long, spiky tongue, whereas the woodpecker finch progenitor experienced a mutation that promoted picking up twigs and poking them into holes. In other words, neither route was superior, and which mutation occurred was just the luck of the draw.

A second possibility is that history matters—that how a species responds to natural selection depends on how it has evolved in the past. Consider the aye-aye, a member of the lemur family. Primates, like all mammals, have mouths composed of bone, skin, muscle, and, usually, teeth.* Evolving a hardy, pointy beak like that of a bird would be a difficult evolutionary feat for a mammal, certainly harder to accomplish genetically than modifying the already present incisors for tunneling into wood. Conversely, birds have modified their forelimbs into flight

---

* A few mammals, mostly ant- and termite-eaters, are toothless.

structures; for them, finger bones are not available to modify into a structure like the aye-aye's cadaverous hook.

SO, WHERE DOES THAT LEAVE US? Is convergence pervasive, a demonstration of inherent structure in the biological world, channeled by predictable forces of natural selection toward outcomes predestined by the environment? Or are examples of convergent evolution the exceptions, cherry-picked illustrations of biological predictability in a haphazard world in which most species have no evolutionary parallels?

We could argue these points back and forth until we're blue in the face. I'd throw out the platypus, you'd counter with convergent hedgehogs; I'd postulate the unique, algae-encrusted, upside-down-hanging tree sloth, you'd retort with bipedal-hopping mice independently evolved on three continents. And that is how, essentially, this controversy has been debated historically, by compiling lists and telling stories.

Conway Morris and his colleagues are to be commended for bringing convergent evolution to the forefront. We all knew about convergence as a neat trick of natural history, a striking example of the power of natural selection. But Conway Morris and company have made clear that evolutionary duplication is much, much more common than we realized. We now recognize that it's a frequent occurrence in the natural world, with examples all around us. Still, it's far from ubiquitous. Seemingly just as often, maybe more often, species living in similar environments don't adapt convergently.

At this point, we need to go beyond documenting the historical pattern, chronicling yet more examples pro and con. Rather, we need to ask whether we can understand why convergence occurs in some cases and not others—what explains the extent to which convergence does or doesn't occur, why bipedal-hopping rodents have evolved independently in deserts around the world, but the kangaroo has only evolved once. And to do that, we need to do more than add additional examples to our lists. We need to test the evolutionary determinism hypothesis directly.

The experimental approach has been the standard for many scientific disciplines for the last century, and for good reason. By carefully altering one variable and keeping others constant, we can directly test cause and effect. Non-experimental studies suffer from the lack of controls and the possibility that any one of many variables may be responsible for the observed differences between study subjects.

Evolutionary biology was late to the experimental game, however— evolution's legendarily languid pace made the idea of experiments a non-starter. We now know that this view is mistaken, that evolution can proceed very quickly. And that realization opens a new door to the study of evolution.

So far, we've been riffling through the drawers of natural history, looking backward through time to make sense of what arose in the past. But now it's time to look forward, to harness the power of the experimental approach to study the evolutionary roles of contingency and determinism.

Part Two

EXPERIMENTS
IN THE WILD

chapter four

The Not-So-Glacial Pace of Evolutionary Change

I t's little appreciated that Charles Darwin was a great experimentalist. At a time when the method was in its scientific infancy, Darwin was setting up experiments to see whether seeds could survive immersion in salt water (some can), how plants grow toward light (the tip of the growing plant is the key), and whether worms respond to music (for the most part, they don't). But Darwin never designed an experiment to test his greatest idea, the theory of evolution by natural selection.

The explanation for this inconsistency is simple: conducting such a study would have seemed pointless. Darwin thought that evolution occurred at glacial speed, so slowly that its progress could only be detected over the course of eons. "We see nothing of these slow changes in progress, until the hand of time has marked the long lapse of ages," he said in *On the Origin of Species*. An evolution experiment would take thousands of years to yield results, much too long to be practical. As far as we know, Darwin never contemplated such a project.

Darwin's scientific track record was pretty amazing. He correctly figured out how coral reefs form, the role that worms play in aerating the soil and, of course, not only that evolution occurs, but that natural selection is its primary engine. So it's not surprising that if Darwin said that evolution moves at a snail's pace, then for more than a hundred years, that's what the field thought.

Of course, in Darwin's time, there were no actual data on the pace of evolution. No one was out studying populations, seeing whether and to what extent they changed through time. Rather, Darwin based his views on conventional wisdom about the pace of geological transformations and Victorian sensibilities about the appropriately low rate of innovation in modern life.

In the last half century, however, we've learned that Darwin got this one wrong. Far from moving imperceptibly slowly, evolution sometimes—perhaps often—moves at light speed. Contrary to what Darwin thought, natural selection can be very strong, and when it is, populations can change substantially in a short period of time.

This sea change in our understanding of the pace of evolutionary change resulted from several different types of data, all of which became available in the middle of the last century. Probably the most influential was the now-famous story of the peppered moth in nineteenth-century Great Britain.

Biston betularia doesn't look like much. A small, grayish-white moth, the size of those that flutter near a porch lamp on summer nights, it gets its name from the little specks of black dotting its wings. Who would have guessed that such an unassuming lepidopteran would become an icon of evolution?

But it did. Two hundred years ago, peppered moths lived up to their name, gray with black drizzle. Sure, the occasional mutation that produced a different color or pattern would come along, but those never lasted long. The reason is simple: peppered moths spend their days resting on trees, their wings outstretched and flat. And trees in the English woods had a peppered countenance as well. Normal moths

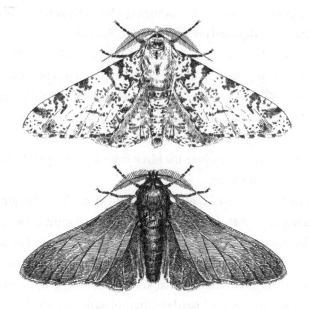

Peppered moth

blended in quite well, and mutants stood out conspicuously, sitting ducks for any keen-eyed bird looking for lunch.

Then, in the middle of the nineteenth century, the Industrial Revolution changed the world for humans and moths alike. For humans, the effects were many and varied, resulting in progress and social upheaval in equal measure. For moths, the results were simple: in industrial centers and areas downwind, trees were covered with the black soot belched from factory furnaces. And for a light-colored moth, that was a problem. Previously supremely camouflaged, they were now easy to spot against blackened tree trunks.

There are few images more comical than a butterfly or moth collector, especially one from the nineteenth century. Picture a gentleman naturalist, wearing baggy trousers, a wool suit and tie, perhaps spectacles and a wool cap. Now picture the same individual running after an undulating lepidopteran, wildly swinging a butterfly net, usually missing as the insect dips and dodges at the last moment.

Back then, butterfly and moth collecting was all the rage—there were meetings and societies and newsletters. And it was big news when something new was discovered. Each novel find was dutifully reported in outlets such as *The Entomologist's Monthly Magazine* and *Entomologist's Record and Journal of Variation*. As a result, we have good documentation of when and where the black color form appeared and how far and how fast it spread.

The first dark-colored moths were captured in central Britain, first in Manchester in 1848, followed by another in Yorkshire in 1860. Soon thereafter, they were detected farther north and south, in London by the end of the century. By the 1950s, dark moths were found throughout much of England. In industrial areas and downwind, the populations were often composed nearly entirely of dark individuals.

Although scientists studied the spread of the black form of the peppered moth since the start of the twentieth century, for a long time the significance of this rapid evolutionary transformation was not widely appreciated. That changed when Bernard Kettlewell, a British physician turned entomologist, conducted a set of now-classic experiments. Kettlewell released both types of moths in forests in rural and industrial areas and later returned to recapture as many of them as possible. His study showed that the survival of dark moths was substantially higher in forests near industrial areas where the trees were darkened by soot. By contrast, in more pristine, rural areas, the typical gray type was more successful. Additional studies demonstrated the agent of selection—by presenting birds with moths placed against different backgrounds, Kettlewell confirmed that birds were adept at picking off moths mismatched against their background.

These studies quickly became *the* textbook example of natural selection operating in nature. The experimental demonstration of strong selection, combined with the historical record of rapid change in coloration over a few decades, clearly indicated that evolution by natural selection could proceed rapidly.*

At about the same time as Kettlewell's work, scientists and the public were becoming aware of rapid evolution in the world around them. Penicillin, the "miracle drug" expected to usher in a future free of infectious disease, was first put into widespread use during World War II. Almost immediately, *Staphylococcus* (the cause of staph infections) evolved resistance and by the mid-1950s, most of the health gains of penicillin had been lost.

Subsequently, as each new antibiotic was developed, bacteria quickly evolved resistance: tetracycline was introduced in 1950 and resistant bacteria appeared nine years later; erythromycin rolled out in 1953—resistance detected in 1968. Methicillin only had two resistance-free years after its 1960 debut. Rapid evolution was very visible to the public and costing human lives.

At the same time that microbes were rendering antibiotics impotent, various pest species were doing the same to our newly developed

---

* The peppered moth story requires not one, but three footnoted points. First, the story of the rise of the dark-colored moth has a pleasingly symmetrical second half. In 1956, England passed the Clean Air Act in response to the London Great Smog of 1952, and very quickly pollution levels dropped, as did the occurrence of dark-colored moths. Today, the dark form has become uncommon throughout Great Britain and has disappeared entirely from some areas.

Second, this story is an excellent example of convergent evolution. The peppered moth doesn't occur just in Great Britain, but throughout the Northern Hemisphere, and the frequency of the dark form has risen and then fallen in other localities as well. The situation has been particularly well documented in North America and the chronicle is almost the same on this side of the pond. The intercontinental convergence goes even one step further: a mutation in the same gene is responsible for dark coloration in moths from both England and the United States.

Finally, in recent years there have been a number of attempts to discredit the peppered moth story, an endeavor loudly cheered on by creationists. It certainly is true that Kettlewell's methods were crude by contemporary standards—the study of natural selection has advanced enormously in the last sixty years. Nonetheless, recent studies by other scientists have resoundingly reaffirmed Kettlewell's findings.

pesticides and herbicides. Field bindweed was the first plant to evolve resistance to an herbicide in 1950 and it was soon joined by many others that became impervious to our chemicals, some almost as quickly as new herbicides were developed.

The same was true for animal pests. DDT was used widely in World War II; the first evidence of resistance was detected in the early 1940s, and by the 1960s, resistance was widespread. Rats evolved resistance to the rodenticide warfarin in 1958, ten years after it was introduced. Overall, the number of insects known to be resistant to some type of insecticide increased from 7 in 1938 to 447 in 1984.* Rapid evolution was costing billions of dollars in damage and, in some places, causing famine and misery.

PEPPERED MOTHS, MICROBES, PESTS. By the middle of the twentieth century, the tide was turning on Darwin's slow-evolution view. But there was still one big catch. Darwin had been talking about evolution in the natural world. And a commonality of all the examples just discussed is that they involved species adapting in response to radical environmental changes caused by humans. Whether confronting air pollution or exposure to initially devastatingly effective drugs, the species in all of these cases faced very strong and novel selection pressures unlike anything they'd previously experienced. Moreover, these strong selection pressures usually were not transient; rather, selection was strong and consistent from one year to the next, unlike how selection was thought to act in nature (remember, there were few data from field studies at this time).

Geneticists since the 1920s had demonstrated that fruit flies and other animals adapt rapidly when exposed to strong and constant se-

---

* By latest count (in 2008), the number now stands at 553.

lective pressures in the laboratory. Similar types of selection lead to the development of new animal breeds and agricultural crops. The peppered moth and resistant microbes and pests could be seen simply as the natural-world analog to laboratory and agricultural studies—we already knew from these artificial selection studies (as they are called) that when humans apply constant selection, populations rapidly adapt. These new examples showed that rapid evolution in response to similar selection pressures happens out in nature, just as it does in the lab or on the farm.

This perceived equivalence suggested that the observed rapid adaptation was not representative of evolution through the eons. In untrammeled nature, it was argued, selection is rarely so strong or consistent. In the natural world, untainted by humans, evolution probably occurs at the much more sedate pace Darwin envisioned. It's only when humans muck things up that evolution goes into overdrive.

Ironically, it was research on the birds bearing Darwin's name—the Galápagos finches—that drove the dagger through the heart of the idea that evolution is always slow. Like the peppered moth, Darwin's finches have become one of the poster-child examples of evolution, and not just because of their name and history. Rather, much of their fame stems from the extraordinary forty-year research program of Princeton biologists Rosemary and Peter Grant.

Starting in 1973, the Grants spent several months each year on the small, crater-shaped Galápageian island of Daphne Major. Their goal was to study the population of the medium ground finch (so named because there are both larger and smaller ground finch species) to see whether and how the population changed from one generation to the next and to attempt to measure natural selection driving such change.*

---

* Actually, this wasn't their goal when they first arrived on the ground, but the project quickly transformed into a long-term study of natural selection and evolutionary change.

To do so, the Grants had to capture and measure all of the finches on the island every year. Only in that way could they see if the characteristics of the population—body mass, beak size, wing length, and so on—were changing from one generation to the next.

Catching a bird is a more passive process than capturing a moth or a lizard. Instead of actively seeking out the quarry and snaring it with some sort of contraption, be it net or noose, ornithologists let birds catch themselves. The trick is to put up what looks like an oversized badminton net, except that the latticework is very thin. So thin, in fact, that a bird often fails to see it until it's too late, flying straight in and becoming hopelessly entangled. Then one of the Grants would come along and carefully extricate the protesting finch, popping it into a cloth sack and bringing it back to camp for processing.

The camp itself wasn't much—a rock shelter, some tarps for shade, and folding chairs. Using calipers, the Grants would carefully measure the dimensions of the beak: how long, how high, how wide. Then they would deftly extend the wing to record its length and measure the leg bones as well. Finally, they would place several colored bands around each bird's leg, giving the bird its own, individual identification card.

Returning year after year, the Grants were able to watch the evolutionary process unfold. Natural selection occurs when the survival and reproductive success of an individual is related to its phenotype. The Grants' data allowed them to ask whether natural selection was operating. Each year, they tabulated who had survived from the previous year and who hadn't. They already had all of their phenotypic measurements for the birds, so they could simply correlate the two data sets: was there a relationship between how long a bird's legs were or how wide its beak was, and whether it survived or perished?

It didn't take the Grants long to find out. The fourth year of their study was extraordinarily dry. Whereas in a normal year, the wet season brings five inches of rain, in 1977 the amount was less than an inch. Daphne Major turned into a barren wasteland. Plants dried up. Water was even more scarce than usual. Seeds—the staple diet of these finches—became few and far between.

The birds died in droves. Starvation and lack of water were a powerful one-two punch, especially because

Medium ground finch eating a large Tribulus *seed*

hungry birds couldn't produce new feathers, and as the old ones wore out, water loss through the exposed skin increased. In January 1977, Daphne Major had 1,200 medium ground finches; twelve months of drought later, the total was 180.

But the mortality was not random. Rather, the largest birds and those with the biggest beaks survived better. The reason is that the supply of small seeds was eaten up first, and as they disappeared, smaller-beaked birds were out of luck—they didn't have the jaw power to crack open the remaining larger seeds. This was among the strongest episodes of natural selection ever detected in the wild.

Natural selection doesn't necessarily lead to evolutionary change. If birds with big beaks survive and reproduce better, then average beak size should increase through time. But this expectation only holds true if big-beaked birds give rise to big-beaked offspring. That is, variation in a trait must have a genetic basis so that trait values are inherited from parent to offspring. Often this is the case, but not always. In humans, for example, the children of bodybuilders don't necessarily have big muscles.

Or think about the sun-loving houseplant in your kitchen window. Put it in a shady corner, and it will grow much more slowly. The amount of water and fertilizer you give it will also matter. Better yet, take a dozen genetically identical plants, created by grafting or taking cuttings, and give them different combinations of light, water, and fertilizer. After a few months, you'll almost surely end up with very different-looking potted plants.

The phenomenon in which genetically identical organisms produce different phenotypes depending on their environmental circumstances is called "phenotypic plasticity." This is the nurture part of the nature-versus-nurture debate.

In the case of Darwin's finches, however, phenotypic variation was genetically based, inherited from parent to offspring. The Grants' team demonstrated that by comparing parents and offspring—they knew whose parents were whose because they had been banding birds shortly after they hatched, while still in the nest. And what they found was a strong correlation between parent and progeny—overall body size was highly heritable, as were beak, wing, and leg dimensions.

Consequently, the larger body size and beaks of the drought survivors were passed on to the next generation, and in subsequent years, the finches were larger and with bigger beaks. Strong selection had led to rapid evolutionary change.

The Grants continued studying the finches of Daphne Major for another thirty-five years. And what they found was that such strong selection was not uncommon. Just a few years later, for example, one of the strongest El Niño events ever brought fifty-four inches of rain—that's right, ten times the normal. The deluge produced a glut of small seeds that led to strong selection for small-beaked birds with the delicate touch necessary to efficiently harvest small seeds. And once again, the population rapidly evolved in response.

The Grants' work was so influential not only for what they documented, but for what they showed was possible. Contrary to conventional wisdom, they demonstrated that evolution can be studied in nature as it occurs, in real time. Their work has been an inspiration to several generations of field biologists, and the result is that the number of people conducting similar research has exploded, providing a previously unavailable wealth of information on rates of evolution in nature.

THE MESSAGE from all of these studies is clear: when the environment changes, species can adapt very quickly. Quickly enough to observe with our eyes. Quickly enough to document during the course of a five-year research grant.

Even just a few years ago, it was big news to document rapid evolutionary change occurring over a short period of time. Now, it's the expectation. Failure to document rapid adaptation has become the exciting find, the unexpected result requiring explanation.

Darwin was a clever experimenter, adept at using the simple materials available to him to test his ideas. For example, in his study of the hearing ability of worms, he observed their reaction to a loud whistle, a bassoon, a piano, and his own shouts. The worms ignored all these auditory insults, but when placed on top of a piano while it was played— instead of on an adjacent table—the worms became quite agitated. Apparently sound is one thing, but vibration quite another.

Given Darwin's experimental proclivities, we can only wonder what he would have done had he known how rapidly evolution can occur. But he didn't know that, so he never devised an experiment to test his theory of evolution by natural selection. And, following Darwin's lead, it was more than a century before scientists gave it a try.

Stephen Jay Gould was an early proponent of the view that evolution sometimes proceeded extremely quickly. His much-debated theory of punctuated equilibrium argued that evolution was episodic, with long periods of little change interrupted by brief bursts of major change. Yet, Gould didn't make the connection between rapid evolution and the possibility of making his replay gedankenexperiment more than an impossible thought.* That job was left to a new generation of scientists.

---

* Probably because he was thinking in geological, rather than human, time scales: rapid over tens of thousands of years does not necessarily equate to rapid over tens of years.

chapter five

Colorful Trinidad

We can't do the sorts of experiments that Gould proposed, turning back the clock millions of years and letting evolution proceed from the same conditions. That was as obvious to Gould twenty-five years ago as it is today—no one has invented a time machine in the interim. But that doesn't mean that Gould's ideas—or the general concept of evolutionary determinism—are immune to the experimental method.

Conway Morris' logic—that convergent evolution embodies the essence of an evolutionary replay recurring across space, rather than repeated in time—can be applied to evolution experiments. Instead of cataloging examples pro and con, researchers can test for convergence directly, thus experimentally testing the replay hypothesis.

Suppose, for example, insects living in lush areas are green, whereas those occurring in dusty, arid areas are brown. By experimentally establishing a population composed of brown individuals in a verdant locality, we can test the hypothesis that the experimental population will converge on the phenotype of the natural populations experienc-

ing the same conditions. Or, researchers can subject multiple populations to similar conditions to test whether they convergently evolve the
same responses. Combining the two approaches is even stronger, testing whether multiple populations evolve the same response exhibited
by populations naturally experiencing the selective condition.

Scientific disciplines are sometimes divided into two categories: experimental and observational sciences. And there's a fair bit of chau-
vinism that accompanies this partitioning, the experimentalists, at
least some of them, considering themselves superior, looking at non-
experimental science as inferior—not even science, according to some
extremists.*

Of course, that view is ignorant. Much can be learned from careful
observation and comparison of natural phenomena, even in the absence of manipulative experimentation. Moreover, experiments have
their own limitations—they're constrained in size and scope. Try doing
an experiment on the cause of volcanic eruptions or the gravity of a
moon, for example.

More importantly, rather than being alternatives, observation
and experimentation are the yin and yang of the natural sciences—
observations from the natural world are the source of the hypotheses
to be tested experimentally. And there's no better example of this than
the research program that inaugurated the era of field experiments in
evolutionary biology.

I'VE BEEN TO RAINFORESTS all over the world, and in many respects,
this one is pretty ordinary: luxuriant vegetation, extreme humidity, the
background din of insect and bird calls. Lots of snakes, all of them fascinating and most non-venomous. Still, I'm glad for my knee-high

---

* "Boy Scout science" being another derogation.

wading boots, just in case my foot lands in the wrong place. What's unusual is how dense the vegetation is, nearly impenetrable. So thick, in fact, that we're actually walking in the stream, the primary reason for my rubber galoshes. And although they're top-of-the-line Bass Pro outdoorsmen's footwear, I'm still moving very carefully because the rocks are extremely slick.*

There's another reason we're walking upstream. Normally when I go to a rainforest, I'm there looking for lizards. But this time my quest is for a tiny fish. Guppies are ubiquitous in homes and classroom fish tanks. They're renowned for their ostentatious flair—bright colors, big spots, oversized tail fins. Like we've done with dogs and pigeons, humans have crafted them into a circus of bizarre varieties and, again like dogs and pigeons, they are the subjects of shows, contests, and commercial ventures.

But just as all breeds of dogs are descended from an ancestral wolf, the variety of guppy breeds are the descendants of a wild species that hails from northern South America, including the island of Trinidad, just off the coast of Venezuela (Trinidad was actually connected to South America when sea levels were lower a few thousand years ago). Unlike the wolf, which shows little hint of the incredible variety of its doggy descendants, wild guppies themselves are a pretty diverse lot, varying greatly in color and ornamentation from one population to the next.

Finally, we get to a small pool. We stand quietly, waiting. After a few moments, the small fish start moving again. A quick swoop of an aquarium net, and we can take a look. It's a guppy all right, but a pretty boring one. Not like the ones you see in pet stores, it's a bland silver-gray, lacking much color or ornamentation.

---

* Slippery enough that after almost falling on my butt several times on the first day, I switched to professional-grade boots with metal cleats on the bottoms.

After a few minutes and several more guppies, we continue sloshing our way upstream, trying to make as little disturbance as possible. The guppies generally live only in quiet pools, so there are long stretches of more rapidly flowing water lacking our quarry. We pass by several more pools with similar-looking fish, but finally we come to one in which the guppies are resplendent in orange and blue, bedecked in black spots and stripes, shimmering in their iridescence, flashy tails trimmed in black.

The contrast repeats itself in tributaries throughout northern Trinidad: dull fish down below, gaudy dazzlers upstream. This is just the sort of observation that excites an evolutionary biologist: repeated, convergent differences indicate that something is driving the divergence in guppy color pattern, but what is it? We now know the answer thanks to a remarkable set of studies spanning more than half a century, research that not only explained what happened in the streams of Trinidad, but made the tiny guppy an evolution celebrity.

The story begins in the middle of the last century with the sort of

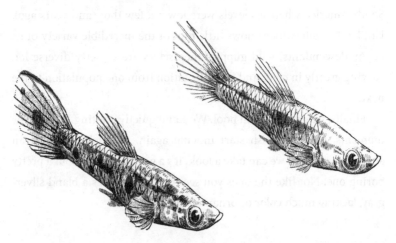

Guppies are more elaborately ornamented in upstream pools (left) than in pools at lower elevations (right).

Renaissance man or woman increasingly rare in this age of specialization. Caryl Parker Haskins published his first paper, on the role of chemistry in agriculture, at age eighteen, followed by several papers on ants while he was an undergraduate at Yale. From there he moved to Harvard, receiving his doctorate in 1935 for research on fruit fly genetics. The career that followed took him in many directions.

He spent time in the laboratories of General Electric, studying the effect of radiation on mold spores. He wrote an acclaimed book, *Of Ants and Men*, comparing ant and human societies.* He was co-owner of an early color photography company and helped develop prosthetic devices to aid soldiers blinded in World War II. Other work involved microbiology, nutrition, and genetics.

All well and good, but where do the guppies fit in? Early in the twentieth century, the Danish scientist Øjvind Winge had established guppies as a good organism for genetic studies, especially for research on the inheritance of sex-linked traits—traits that only occur in one sex.† Haskins picked up on this work, and starting in 1932 he began his own guppy-breeding projects to understand the inheritance of color traits, which occur primarily in males.

The guppy genetics work of Haskins and others advanced rapidly, leading to many inferences about how the guppy's genetic system had evolved. Haskins realized, however, that without more information from natural guppy populations, these ideas could not be tested. So in

---

* He told the *New York World Telegram* in an article after the book was published, "They are fascinating creatures. When they go to war they stab each other, spray poison and cut off each other's heads. They subjugate weak races and keep slaves. But they can be kind, too. Nothing they like to have around more than a pet beetle," and, "Like people, they're the most adaptive organisms in their group and, similarly, their most dangerous enemies are others of their own kind. Like people, they go anywhere."

† Winge's work was not restricted to fish. He worked on a wide variety of organisms, but is best known as the father of yeast genetics, and thus the intellectual ancestor of some of the laboratory experiments discussed in Chapters Nine, Ten, and Eleven.

*A typical Trinidadian stream waterfall with a pool full of guppies
and, probably, predators*

1946, he and his scientific collaborator and wife, Edna Haskins, set out
to examine the distribution of guppy color patterns and their underlying genetic basis in the wild.

Haskins quickly noted a topographic feature that affected some fish
species more than others. In the mountains of northern Trinidad, some
streams were punctuated by waterfalls, most a few feet in height, but
some plummeting as much as thirty feet. These cataracts formed a barrier to most fish, including the guppy's major predators, but two fish
species were commonly found above these obstructions, guppies and
the Trinidadian killifish.

Salmon athletically jump over cascading falls to make their way upstream. But guppies and killifish are too small and the plunge too high.
Instead, these fish do the same thing that canoeists do when facing an
upstream obstacle—they portage around it. Like a few other fish, killifish are renowned for their ability to wriggle great distances across
damp forest floors; they simply crawl out of the pool at the bottom of a

waterfall and scrabble up the hillside to the pool above. Guppies are not quite so pedestrian, but they are able to swim in very shallow and temporary water bodies. Although it has never been directly observed, the fish most likely make their way upstream through temporary channels that form on the forest floor during periods of heavy rains.

The evolutionary significance of this waterfall barrier—permeable to some fish species and not others—became clear once Haskins compared fish from different parts of the Trinidadian streams. In the low-predation populations above waterfalls, the males, but not the females, were very colorful, whereas in the high-predation populations lower down, both sexes were quite drab. Haskins came to the conclusion that the presence or absence of the major fish predators on guppies was the key factor affecting the evolution of male color.

To test this idea, Haskins conducted lab predation studies, putting guppies in aquaria and outdoor pools with a variety of Trinidadian predators. Sure enough, brighter fish disappeared from the tanks at a much higher rate than drabber ones, confirming that being colorful was a big disadvantage in predator-filled waters.

But why were males colorful in the absence of predators? That's where the ladies come in. For reasons we still don't fully understand, female guppies prefer flashy males, and this preference is stronger in predator-free areas. This knowledge has been gained in a series of elegant studies over the last twenty years, following a line of research initiated by Haskins in another of his aquarium studies.

Haskins' guppy research was extremely innovative and insightful, all the more impressive because it was a sideline to his many other projects and responsibilities. Indeed, this work was so subsidiary that it isn't even mentioned in the mini-biographies and obituaries published after his death. Nonetheless, the work set the stage for one of the most exciting research programs in evolutionary biology in the late twentieth and early twenty-first centuries.

IN THE HEADY DAYS of the 1960s, at least one student at the University of California, Berkeley, was paying attention to his schoolwork. Having spent his childhood days chasing lizards and catching bugs in the chaparral of Southern California, the young John Endler went to Berkeley to major in zoology. Once there, he quickly gravitated to the university's famous Museum of Vertebrate Zoology, jam-packed with specimens and scientists studying them (and, incidentally, quiet; Endler noted that "all that protest stuff took place at the opposite end of campus, so I wasn't bothered"). Most people think of natural history museums in terms of their great public exhibits, such as the fabulous dioramas full of stuffed animals at the Smithsonian and other museums. What many don't know, however, is that behind the scenes, many museums house large collections of preserved plants and animals, carefully cataloged and curated, available for scholarly study by their own staff scientists as well as by visitors from elsewhere.

Endler's duties as an assistant to the curator of herpetology included working with the pickled reptiles and amphibians in the collection. As he did so, he took to examining specimens from different regions and learned that they often are highly variable from one place to another. Such geographic variation is a well-known phenomenon in the natural world. Human skin color and facial shape are classic examples of how populations often diverge across a species' range.

In addition to learning about geographic variation, Endler slipped another trick up his sleeve while at Berkeley: he learned to devise experiments to test the ideas he developed. One experiment involved spinning walking newts around in a can to see if the rotations would disrupt their ability to navigate by the stars (it did, leading to his first publication in a scientific journal). Another involved fitting lizards with aluminum hats to control the amount of light hitting their heads in an attempt to test hypotheses about how their daily rhythms are regulated (alas, foiled by equipment malfunctions).

Fascinated by the phenomenon of geographic variation, Endler went off to Edinburgh to conduct doctoral studies with an expert on snail diversity. There he developed theoretical ideas about how geographic differentiation could eventually lead one widespread species to break into multiple, non-interbreeding species.

With his experimental mindset, however, Endler was not satisfied with simply developing a provocative theory about how new species arise. Rather, he immediately came up with an experimental way to test the idea. The result was an elaborate laboratory study on fruit flies in which populations were established in different cages. These populations were subjected to different selection pressures, mimicking a situation in which the populations lived in places that favored different traits. Each generation, some of the flies were exchanged between populations, simulating the dispersal that naturally occurs among adjacent populations. Traditional theory predicted that this swapping of individuals would have the effect of homogenizing the gene pools of adjacent populations. The results of the experiment, however, contradicted this theory. Even in the face of such genetic sharing, the different selection forces acting on the populations caused the populations to diverge genetically, exactly what Endler's theory predicted.

This work was a great success and led to the publication of several important papers and a widely read monograph that established Endler as an up-and-coming star in the field of evolutionary biology. Nonetheless, at heart Endler was a naturalist, fascinated with the biosphere and what makes it tick. Consequently, even as he labored over his theoretical derivations and laboratory manipulations, he was already thinking about his next step—looking for a project that would take him out into the field.

While reading up on geographic variation for his doctoral work,

Endler had come across the papers on guppy color variation. Intrigued, he wrote Haskins, who by that time (the early 1970s) had become president of the Carnegie Institution of Washington.* More than ten years into his executive role, Haskins was still very much engaged in guppy research, and a lively correspondence ensued. Endler even met with Haskins at the institution's massive headquarters in Washington, D.C., a visit that ended with a tour of the large guppy collection that Haskins maintained in his home. Endler was sold on guppies—the fish would become the focus for his next research project.

Returning to the United States to a faculty position at Princeton, Endler set out to more rigorously investigate Haskins' idea that predation was a key factor in shaping male guppy color. Every year, Endler would spend the entire summer in Trinidad, tromping along streams, charting guppy distributions.

The northern end of Trinidad is bordered by a chain of mountains; cascading down these mountains, on both the north and south slopes, are innumerable streams. Actually, they can be enumerated, and Endler had a map indicating all of them. He set out to visit each stream in areas with intact forest to see which fish species occurred there and what the guppies looked like. In all, he visited 113 sites in fifty-three streams over a five-year period.

Collecting data in these streams was no mean feat. Based on his maps, Endler had an idea of where he needed to go, but getting there was the trick. Sometimes a road ran nearby, entailing only a short hike through the woods. On other occasions, a longer trek—at times better described as a bushwhack—was required.

Once he got to the stream, Endler still had to find the guppies. So he'd walk up- or downstream, looking for a quiet pool. And then the

---

* Now the Carnegie Institution for Science, a prestigious private scientific research organization founded by Andrew Carnegie in 1902.

pool had to have a good population of guppies—some pools are more guppyful than others. When the banks were too steep, he'd walk through the stream, which muddied the water and disturbed the fish, just as I did many years later. At times, the only route was a precarious balancing act along a fallen log. Waterfalls posed a particular problem. The boulders around them, being in the spray zone, were often very slick, making for treacherous footing. And getting above the waterfalls, where often an idyllic, fish-filled pool awaited, could be a challenge, especially if no dead trees had conveniently dropped against the vertical rock walls in just the right spot to aid in the ascent.

Once a suitable pool was located, Endler would quietly sit at the edge for an hour looking into the clear Trinidadian stream waters, noting all predatory fish species and estimating the population density of the guppies. Then came time for the guppy roundup. With a butterfly net in each hand, Endler would patiently herd the guppies and then, with a sudden swoop, he'd net the entire school. At each site, he'd nab about two hundred fish, enough to get about fifty adult males. Each male was examined and the location and color of each of its spots recorded.

It took four years of patient data collection, but the results were worth it. The pattern was clear-cut. Haskins had been absolutely correct: the coloration of male guppies was strongly correlated with predator presence. At sites with few or no predators, such as streams above waterfalls, male guppies were more flamboyant, possessing more and larger spots. Not all spots are the same, however; it was mostly the red and black spots that got larger, while the blue and iridescent ones became more numerous.

These comparisons among sites were strongly suggestive, but not conclusive. As always, correlation is not causation. Perhaps some other factor that was correlated with predator presence could be responsible. For example, the sites lacking fish predators, being above waterfalls, tended to be at higher elevations, where the size of the pebbles in the streambeds was greater than at lower elevation sites. So guppy spot size was correlated not only with the presence or absence of predators, but also with the size of pebbles in the streambed. To the extent that spots helped guppies blend into the background, one might expect spot size to correlate with local pebble size, providing an alternative explanation for the large-spotted fish above the waterfalls (this would imply that something else besides fish was an important predator on guppies—maybe birds).

Endler knew how to solve problems of this sort, explaining in his now-classic 1980 paper that "the field results are striking, but it is possible that some other factor in the environment is affecting the color patterns. In order to provide a more direct test of the hypothesis that the entire color pattern is subject to natural selection, two experiments were set up, one in a greenhouse and one in the field."

Part one was a laboratory evolution experiment. When we think of lab experiments, we tend to think of vials of fruit flies or petri dishes packed with microbes. Indeed, as we'll see later, many lab evolution experiments of just this sort are now ongoing. But Endler's laboratory experiment was on an entirely different scale. No compact stack of petri dishes or rack of vials abuzz with fruit flies here; a population of guppies, particularly one in a naturalistic setting, requires considerably more room than in a standard science lab. Fortunately for Endler, the perfect place for such an experiment was available next door, an abandoned greenhouse. Endler took it over and repurposed it for his ichthyological investigations.

The greenhouse was large, sixty feet long by twenty-five feet wide, full of long tables and botanical paraphernalia. Endler removed everything and set to work. He began by contouring concrete on the floor— doing much of the work himself—to craft natural-looking guppy habitats, complete with pools and streams, each three hundred square feet in area with its own waterfall. In total, he constructed ten enclosures, arranged in three rows separated by two walkways. The pool floors were layered with shockingly colorful aquarium gravel. Unexpectedly gaudy in a New Jersey hothouse, the colors were actually a good approximation of Trinidadian streambeds. Plants and invertebrates transplanted from a local stream created a working ecosystem. Algae, pleased by the ample heat and light, flourished, establishing the basis for a functioning food web.

To populate the pools, Endler mixed guppies from eleven different streams, allowed them to mingle and reproduce for several generations, then placed two hundred fish into each enclosure. Because the fish were randomly assigned, the populations initially differed little in their degree of ornamentation.

The goal of the experiment was to test the predictability of guppy color evolution in relation to the presence or absence of fish predators. Four weeks after placing the guppies in the pools, Endler added predatory fish to some of the pools. Four pools received a highly predatory pike cichlid, a streamlined torpedo with teeth that makes its living eating guppies; four received the much less threatening killifish, which only occasionally eats young guppies; and two pools were left with no predators at all. Endler predicted that if the differences observed among Trinidadian streams were the result of predator complement, then his

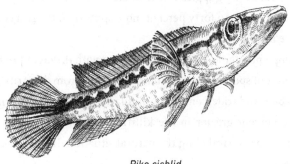

Pike cichlid

populations would diverge in color over time, those cohabiting with pike cichlids becoming drabber, those facing little or no predation brightening up as sexual selection worked its wiles.

Plop! In went the predators, and that was Endler's last intervention in the experiment. Other than the daily feeding and monitoring of the water chemistry, the guppies and their predators were on their own. Guppies have a famously rapid life cycle, capable of breeding when less than two months old. How many generations would it take for evolutionary change to occur? Endler went back to his fieldwork in Trinidad and let nature—greenhouse-style—take its course.

Five months after the predators were introduced, Endler decided to see what was happening. Every fish in each pool was netted and photographed, its spots counted and measured, its color noted. Then the fish were tossed back into the drink, no worse for the wear, to keep on evolving. Five months is at most a couple of guppy generations, surely too little time for evolutionary change.

Wrong! The populations had already begun to diverge. The guppies living with the highly predatory pike exhibited a ten percent reduction in spot number, those with no predator or the inoffensive killifish had increased by about an equal amount.

Nine months later, when Endler next censused the fish, the populations had continued to evolve in opposite directions and the differences were even greater: the guppies in the pools with killifish or no predators now had more than forty percent more spots than the fish cohabiting with the cichlids.

Like guppy populations in Trinidad living with different predators, the divergence in spot numbers primarily resulted from disparity in the number of blue and iridescent spots. Similarly, the size of the spots was nearly fifty percent greater in the killifish-pool males than in those with cichlids, again paralleling the natural situation. Indeed, the similarity between the greenhouse pools and their natural counterparts was extremely high—in only two years, the experimental populations had pretty much converged on guppies living in comparable predation environments in nature.

Impressive as these results were, they still were the results of a laboratory (or at least, a greenhouse) experiment, with all the artificialities that implies: the fish were fed on a daily basis; aside from the guppies and the predators, all other members of the Trinidad ecosystem—birds, other fish, crayfish—were absent; it never rained. If this experiment had been the entire project, the results would have still been considered a remarkable demonstration of rapid evolution in an animal bigger than a fly or a microbe. But the results also would have been discounted, at least by some, as an artifact of the laboratory, not applicable to the natural world.

Endler took care of this objection. Even as he was pouring concrete and building waterfalls, he also had his eyes on a pristine pool in the upper reaches of Trinidad's Aripo River. Just above a waterfall, the pool had killifish, but seemed to lack guppies. He went back repeatedly for two years, diligently searching for signs of guppy life, but there were none. Finally convinced that the pool was devoid of guppies, he initiated part two of his project in July 1976.

Nearby, on other tributaries of the Aripo, Endler found two very similar streams, both with guppies but differing in their roster of predators. One had only killifish, but the other was chockful of guppy-eating predators, including the pike cichlid. And just like elsewhere on the island, the guppies in these two streams differed as expected, being brighter, with more and bigger spots, in the killifish-only stream.

Endler's experiment was simple. He took two hundred guppies from the stream full of predators and placed the drab and small-spotted fish above the waterfall in his pristine, guppyless pool that had only killifish. His prediction, derived from observations of natural populations, was that they would evolve to look like the colorful guppies in the nearby stream with only killifish.

That Endler even conducted this experiment is a testament to his prescience given the state of the field at the time. Just because guppies in Trinidad differ in coloration in alternative environments doesn't mean that those differences evolved quickly. Although the clouds of the impending revolution were already forming, at that time the idea that evolution generally proceeded quite slowly was still the standard wisdom. But Endler did not buy into the dogma. His familiarity with evolutionary theory suggested that evolution ought to be able to occur quite rapidly if natural selection were strong enough, and he was convinced by examples like the peppered moth and resistant pests and microbes. No one had ever set up an experimental study of evolution in the field, but Endler had the audacity to think that an evolution experiment might work.

Two years later, Endler returned, caught up the guppies in the stream, and took the standard measurements. And just as in the greenhouse, natural selection had worked its evolutionary magic: in a few short generations, the guppy population had evolved more and larger spots, transforming from a typical high-predation guppy population to one typical of sites with only killifish.

Endler had shown that guppy evolution was predictable. Not only can we understand why some guppies are colorful and others aren't, but if we re-create the selective conditions—in the lab or in the field— they will evolve exactly as expected.

The broader message to evolutionary biologists, however, transcended guppies: when natural selection is strong, evolution can occur rapidly. And the corollary: the experimental method—the powerhouse of modern science—can be used to study evolution not only in the controlled, but artificial, confines of the lab, but also out in the messy, uncontrolled, rambunctious natural world.

ENDLER'S WORK had an important impact on me early in my studies. As a sophomore in college, I took a small-group class in which we read his monograph on how new species arise. The book opened my eyes to the interplay of theory and data: how observations from the natural world lead to the development of theoretical ideas, which in turn are tested by collecting new data.

Even more influential was a talk I heard him give in the weekly departmental seminar series. By this time, biology nerd that I had become, I was hanging around the graduate student offices, picking up pearls of wisdom and gossip and trying not to make too much of a nuisance of myself. Like graduate students everywhere then and now, this group was a jaded lot, always quick to spot the untested assumptions and outright flaws in a paper or presentation. But not this time. Endler gave a lecture on his experimental work and I remember very clearly, thirty-five years later, walking out of the room and hearing the effusive comments of the grad students. The words of one stand out vividly in my memory: "Who knew evolutionary ecology could be an experimental science?" I had no way of knowing it at the time, but those words and the revelation to which they referred would play an important part in my own career a half dozen years later.

The graduate students and I were not the only ones blown away by an Endler talk. Several years earlier, before the work was even published, Endler gave a presentation on the guppy experiments at the Academy of Natural Sciences in Philadelphia.

In the audience that day was David Reznick, a fourth-year graduate student at the University of Pennsylvania. Like Endler, Reznick had been a boy naturalist, particularly taken with reptiles.* A summer internship during college in the American Southwest had introduced him to lizards that live on lava flows and blend in by dint of their black coloration, quite different from the light color of their relatives that occur on the sand in the adjacent desert. Because the lava is from eruptions that occurred only a few thousand years ago, the populations must have evolved their dark coloration in the recent past. This observation had convinced Reznick that evolution could occur very quickly as populations adapted to new circumstances.

Switching to fish because they are in many ways more practical to work with than lizards, Reznick decided to focus his graduate work on the evolution of life history. The term "life history" refers to all the factors that affect an individual's reproductive success: how long it lives, how quickly it matures, how many offspring it produces per breeding event, and so on ("demography" is an older term with a similar meaning).

A considerable body of theory predicted how life history should vary depending on different conditions. When predation levels are high, for example, individuals should live fast and die young—given the likelihood that they won't live to a ripe old age, they should mature quickly, putting their energy into producing early and often, rather than investing in growing to a larger size. Moreover, because of the predation threat, individuals should spread their bets, producing many small offspring instead of fewer large ones.

On the other hand, when predation levels are low and average life

---

* In this and a number of other ways, Reznick and I are peas in a pod. I, too, was a boy lizard chaser, and I learned during writing this book that we were both influenced growing up by attending the same summer camp, the Prairie Trek, which took teenage kids out camping throughout the southwestern United States. Subsequently, both of us spent time at Washington University in St. Louis, he as an undergraduate, I years later as a professor. Perhaps most strikingly, we've both had our work ridiculed by the *National Enquirer*, Reznick in an article entitled "Uncle Sam Wastes $97,000 to Learn How Old Guppies Are When They Die," me in the piece "Leapin' Lizards! $60,000 of Your Taxes Is Wasted Studying Why They Have Favorite Islands."

span long, individuals have the luxury of investing energy into growth, postponing reproduction until later in life, when, because of their larger size, they can produce many more offspring. Moreover, because offspring have a good chance of surviving, parents should invest a lot in each one, priming them for a life competing with others.

Theories about the evolution of life history had been tested experimentally in the lab with fruit flies and other organisms, but they had never been investigated with experiments in nature. Like Endler, Reznick's understanding of theory suggested that evolution should occur rapidly when natural selection is strong; his own observations of divergent populations in geologically new environments provided supporting evidence.

To test the idea that life history would evolve quickly and adaptively, Reznick started studying populations of the mosquitofish—a bland-looking relative of the guppy—in the marshes of Cape May, New Jersey. But the work was not all that it was cracked up to be. The fish were strongly affected by the seasons, a wrinkle that complicated his comparisons of populations living with or without predators. Moreover, the long commute from Philadelphia, and especially the infamous sights and smells of New Jersey, were not how Reznick had envisioned fieldwork. The tropics were really where he wanted to be.

And then he heard Endler's talk, and all became clear. Who cared if he had invested two years in mosquitofish? The guppies had everything he wanted in a research project, and having kept them in aquaria as a boy, he knew he could work with them.

At the post-seminar dinner that evening, Reznick told Endler about his ideas. A week later the two met down the road in Princeton, where Endler invited Reznick to join him in Trinidad the following year. Reznick's interests in life history perfectly complemented Endler's and Haskins' focus on color.

The next March, Reznick and Endler rendezvoused at a field research station in the mountains of Trinidad. Endler laid his topographic map of the northern mountains on a table and—this being back before the days of portable computers—placed tracing paper on top of it and made a copy. Reznick still has this hand-sketched map—now a prized possession—in his office. On it, Endler's notations indicate which streams would be worth surveying to make a comparison of guppy populations with and without predators. Reznick hopped into a rental car and ventured farther into the mountains. Beginning what has become a nearly forty-year research program, he sampled far and wide, high and low, catching guppies to determine their schedule of life events.

Reznick surveyed sixteen populations that first summer. Just as with color, the demographic differences were marked: guppies in populations with pike cichlids and other predators differed greatly from guppies in the low-predation streams. In particular, high-predation guppies matured at a smaller size, devoted more resources to reproduction, and produced more and smaller offspring. Exactly what theory had predicted! Of course, the underlying assumption is that guppies in the high-predation sites actually live shorter lives. Reznick confirmed that, too, by capturing guppies, marking them, and returning later to see how long the marked fish survived. Over just a two-week period, fifteen percent more guppies died at the high-predation sites than in the low-predation streams; a longer mark-recapture study showed that survival over a seven-month period was about one percent in high-predation sites, whereas in sites with only the killifish, survival was twenty-five times higher.

The pattern of life history differences was compelling. But, like Endler, Reznick saw these results not as definitive conclusions, but as the source of hypotheses, particularly that predation was responsible for the differences among populations. What had really drawn Reznick to guppies was the ability to test ideas experimentally, so he was keen to follow up on Endler's studies. He revisited Endler's Trinidad introduction site and found that the guppies there—descended from inhabitants of a high-predation pool, but placed in a benign environment—had evolved the low-predation life history. Ditto for the greenhouse experiments: two and a half years after Endler began the study, Reznick compared the life histories of the guppies in low- and high-predation enclosures and found differences paralleling those seen in the wild.

Reznick went on to conduct his own experiments, introducing guppies into low-predation streams at two other sites. There, too, the results were very much the same. Finally, he tried a new type of experiment, the converse of those done previously. Instead of moving guppies, he moved their predators, putting pike cichlids into a pool above a waterfall where previously only guppies and killifish had occurred. The cichlids knew a good thing when they saw it and the naïve residents paid the price. Quickly, the guppies began to evolve the characteristics of high-predation populations. Five years into this experiment (the last time Reznick checked), the formerly low-predation guppies displayed life history traits almost exactly intermediate between nearby low- and high-predation populations.

Overall, the results of Endler's and Reznick's studies were strikingly similar. Like Endler's work with color, Reznick's experimental populations evolved just as would be predicted from knowledge of variation among natural populations. Guppy life history appears to be an extremely malleable evolutionary trait, one in which natural selection quickly and predictably molds an evolutionary response.

TO MAKE THE CASE for rapid adaptive evolution airtight, Reznick had to deal with an additional issue: perhaps the differences he was observing between guppies in high- and low-predation environments were not the outcome of evolution. Rather than stemming from genetic changes, the life history differences between populations could, in theory, be the result of environmental influences causing genetically similar guppies to grow and reproduce in different ways, the phenomenon of phenotypic plasticity discussed in Chapter Four.

Reznick addressed that question directly by bringing fish from different populations back to the lab and maintaining them in identical aquaria, all lacking predators. The "livestock"—as he refers to them— were housed separately and allowed to reproduce, and all the babies were raised individually in identical circumstances. This setting is referred to as a "common garden" experiment, named after its botanical roots.

Reznick's goal was to find out whether the differences among females captured in the wild would be passed on to their offspring, all of which were raised under identical conditions. If the differences among the mothers were genetically based, then their offspring should also be different. Conversely, if the mothers' differences had been induced by the environment in which they had grown up, then their offspring, raised in a common garden, should be similar.

The results were as clear as a Trinidad stream. Laboratory-raised guppies and grandguppies still exhibited the differences characteristic of their wild-caught mothers and grandmothers. Those that descended from high-predation streams grew fast and reproduced copiously at a young age, just like their forebears; those that originated from a more placid, low-predation homeland exhibited the laid-back, slow-maturing life history. Reznick thus concluded that the differences must be genetically based, the result of evolutionary divergence.

In a way, Reznick's work was a follow-up to genetic studies conducted by Haskins decades earlier. Haskins had been concerned with whether color differences were genetically based. He took a different approach than Reznick, mating individuals with different phenotypes to examine how traits were passed from parent to offspring, just as

Mendel had done with his famous peas. And like Reznick's study on life history differences, Haskins' work and subsequent studies firmly established that color variation in guppies is also primarily determined by genetic differences.

Scientists are now able to take the genetic study of phenotypic differences to new levels by sequencing the entire genome of many individuals. Reznick's group is currently taking just this approach to try to identify the actual DNA differences responsible for variation in life history, color, and other guppy traits.

MY DESCRIPTION SO FAR has painted the picture of the scientific value of Reznick's work, but it hasn't done justice to what it's like to conduct cutting-edge experimental research in the middle of a tropical jungle. Ever since that first field trip in 1978, Reznick has returned to Trinidad nearly annually and sometimes as many as four times in a year. Some of the work has involved following up on the experimental introductions, and even starting several new experimental populations—most recently four introductions, two in 2008 and two in 2009. But most of the work has been comparative, contrasting the ways in which natural populations have adapted to living in high- or low-predation environments.

Many of these comparisons involve visiting closely situated populations separated by inhospitable rapids or waterfalls. Collecting data on these populations requires spending the day in the forest, hiking from one site to the next, taking measurements, catching fish amidst the tropical splendor.

Brilliant blue butterflies floating by, lizards scuttling in the leaves, beautiful birds in the trees. Frogs croaking harmoniously, drowned out at times by the surprisingly pleasing insect buzz. David Reznick must have the best job in the world.

And yet, idyllic as it may seem, working in Trinidad is not without its dangers, and in his forty years there, Reznick has experienced his fair share. The easiest way to move through the forest, he has learned, is to follow animal trails. Unfortunately, some locals who like to illicitly feast on the forest's denizens also make use of those trails. To acquire their meat, the poachers set up booby-trapped homemade guns, composed of a pipe loaded with a shotgun shell and triggered by a trip wire. These trap guns (as they are called) are set low to the ground, perfect to blow away a tawny, jackrabbit-sized rodent called an agouti or other four-legged fare such as a small deer. But anything that walks by will set them off, including a two-legged biologist in a hurry to get from one stream to the next. Reznick actually was lucky, most of the shot passing between his legs, but seventeen pellets remain in his left ankle and the hearing in his right ear was destroyed by the blast.

Another time a walk along a stream above a waterfall turned into a cliff-hanger. Not just because it was suspenseful (which it was), but because he ended up hanging over a cliff, clutching a bush, Indiana Jones–style. He found himself in this precarious position as a result of slipping on a slick rock and being carried by the rushing water to the edge of the twenty-foot drop, only saving himself by grabbing a shrub at the last possible moment. To make things worse, his only companion had just badly gashed his arm (they were headed to the medic when Reznick lost his footing) and was unable to provide assistance. Fortunately, like so many adventure heroes before him, Reznick was able to muster the strength to pull himself back from the brink, living to experiment another day.

Many of Reznick's misadventures involved snakes. Despite becoming an ichthyologist, Reznick's passion for creepy-crawlies has always remained; routinely after a hard day in the field, he still heads out at night to look for frogs, snakes, or whatever might be found.

In the acknowledgments section of a scientific paper by another group, the authors thanked Reznick for his "sage guidance in the field (sometimes dispensed calmly while standing on a fer-de-lance)." The fer-de-lance is an extremely venomous viper. Reznick claims that the story is exaggerated, but it is true that whereas others give the snakes a wide berth, he always goes in for a closer look, sometimes capturing them for relocation to a more out-of-the-way place.

And then there are the army ants, the voracious hordes that move through an area in the hundreds of thousands, devouring any insect or soft-bodied animal not quick enough to get out of the reach of their sharp-toothed jaws. Humans are not at risk of being consumed, but a tangle with army ants is no fun. A provoked colony goes into kamikaze defense mode; their bites are particularly painful because their long jaws are left embedded in the skin, and they've got a nasty stinger on their rear end as well.

Periodically, an advancing column decides to move straight through the field laboratory building. This is not actually a problem—just push the chairs out of the way, make sure that nothing you need is on the other side of their path, and wait for them to move on. Indeed, their services in clearing the floor of debris and other insects can be much appreciated.

But there was the time that Reznick spied a gorgeous, six-foot-long, gunmetal-blue snake with an orange belly, the machete savane, moving on the forest floor. Diving, he grabbed the snake by the tail, only to have it turn to attempt to bite him—machete savanes may not be venomous, but they do have sharp teeth. Only then did both snake and Reznick realize that they were in the midst of an army ant column and about to be swarmed. Declaring a truce, Reznick dropped the tail, the snake aborted its strike, and they fled in opposite directions, Reznick straight into a nearby stream to dislodge his formic foes. Despite various stings and bites, the ants may have saved Reznick from a worse outcome—on another occasion, a machete savane that he grabbed by the back end responded by turning around and biting him on the nose.

Finally there are the flash floods. Much of the fieldwork is done in the wet season, when a thunderstorm can pop up at any time. The streams in which they work are often in narrow ravines, and when a storm occurs upstream, a flash flood can come barreling down with no advance warning. Reznick and team have had several close calls, but no one has ever been injured.

REZNICK AND OTHERS have now surveyed a large number of sites, and the life history of the guppies in these populations can be accurately predicted by knowledge of whether guppy predators are present. Given this consistency among natural sites, we would expect that the results of experimental translocations would be predictable, and indeed they are.

Haskins' and Endler's surveys showed a similar predictability with regard to color: male guppies are more garish in sites at which predation pressure is weak. Endler's experiments—the ten greenhouse ponds and the single natural introduction—produced results consistent with each other and with natural patterns of variation. Yet, surprisingly, given the renown of Endler's work, no one examined color evolution in any of the other guppy translocations for the next quarter century. Finally, in 2005 Reznick and Endler teamed up with experts in animal vision to look at the fish from one of Reznick's introductions.

Like the fish in Endler's introduction, the Reznick guppies also had become more colorful when moved from a high-predation pool to one containing only killifish. However, the way they had done so differed in the two populations. The Endlerians had become more vibrant by increasing the amount of all colors. By contrast, Team Reznick had become substantially more iridescent, but the size of the red and black spots had not increased—in fact, the red spots had actually gotten smaller, perhaps to make room for more iridescence.

Why the different ways of becoming more ostentatious? Beside the presence of predators, many other factors affect guppy color. The best color—either for camouflage or flamboyance—depends on how much light filters down through the canopy and how murky the water is. Even the size of the rocks on the streambed may play a role. In his greenhouse experiment, Endler varied the size of pebbles—large or small—and found that in the presence of predators (but not in their absence), spot size evolved to match the background.

Alternatively, the different colors might have nothing to do with differences in the environment, and might, instead, simply be a result of historical contingency, the fact that the two populations had different evolutionary histories. The ancestral Reznick population is much more iridescent than the ancestral Endler population. We don't know why that is, but we do know that females choose their mates based on color, and females in different populations have different preferences. It may well be that females from the ancestral Reznick population have an inordinate fondness for iridescent males; in the absence of predators, this female preference may have driven the evolution of increased male iridescence.

But that's just a guess, a hypothesis awaiting further study. At this point, all we can say is that moving fish to low-predation environments leads to increased ornamentation, but predicting what form this ornamentation will take is not possible.

RESEARCHERS HAVE USED guppy experiments to examine the predictability of evolution of one other trait. Anne Magurran, an Oxford expert on guppy behavior, realized the opportunity presented by the guppy experiments. Like so many other aspects of guppy biology, behavior differs between high- and low-predation populations. In the presence of predators, guppies from high-predation localities tend to stay in groups and keep their distance from the predatory threat. By contrast, guppies in low-predation environments have lost their wariness: they seek the protection of the herd much less and approach the predator more closely. Would guppies from experimental populations—moved just a few years before from a high-predation to a low-predation site—behave like the wary guppies in their ancestral homeland? Or would they have evolved the more carefree lifestyle of the non-predated?

To find out, Magurran brought fish into the lab and bred them, raising the offspring in the absence of predators. Then she conducted behavior experiments in aquaria, placing guppies with a school of fish and a realistic model of a predator to see how they responded.

The results of the laboratory behavior experiments were unambiguous: the fish had the devil-may-care behavior of the unconcerned. Put them with a school of fish, and they forsook the safety of the herd, instead hanging out by themselves; put a model of a pike cichlid in the tank and they swam right up to it to take a look. Just like life history and color, guppy behavior evolves rapidly and predictably in experimental introductions.

Shyril O'Steen, an Olympic gold medalist turned evolutionary biologist,* took this work one step further by looking not at the guppies' behavior, but at the outcome of their interaction with predators. Have guppies coexisting with predators evolved greater escape abilities than guppies living the good life in the absence of predators? To find out, she collected guppies from three experimental populations. Two were populations introduced to low-predation pools by Endler and Reznick, and the third was the population from the site where Reznick had introduced pike cichlids. For each of these experimental populations, O'Steen sampled another population to use as a comparison. For the

---

* O'Steen won a gold medal in the 1984 Olympics as a member of the U.S. women's coxed eight rowing team.

two guppy introductions, she used their ancestral populations, and for the pike cichlid introduction, she used a nearby population that remained cichlid-free.

To test her hypothesis, O'Steen conducted predation trials in the laboratory, placing guppies from a pair of populations in a pool with a pike cichlid. As O'Steen predicted, guppies from cichlid-free pools had much poorer survival skills than their predator-savvy counterparts: in all three comparisons, naïve guppies were eaten at twice the rate of their savvy counterparts. Subsequent studies indicated that guppies that occur with predators are not only warier, but also better at escaping once attacked. Common garden studies confirmed that the differences are the result of predictable evolutionary change.

THIS IS A GOOD PLACE to discuss the ethics of experimental introductions. Invasive species are a major economic and environmental problem the world over. A blanket prohibition on intentional introductions of species to places where they don't naturally occur, even for scientific purposes, would seem reasonable to many. Indeed, Magurran and her colleagues have called for just such a moratorium on future guppy introductions.

Several potential harms may result from moving fish around. First, such introductions disturb the natural order. Guppyless pools are the result of natural processes, and their occupants have adapted to living without guppies. As current work by Reznick and others is showing, adding guppies to a pool causes major changes to the ecosystem. In this regard, one might view such introduced guppies as just another invasive species disrupting the natural order, no different than the brown tree snakes that ate the birds of Guam to extinction, or water-sucking tamarisk trees that are transforming the parched American Southwest.

Moreover, the impact of guppy introductions is not limited to the pool into which they are released. Guppies sometimes may be stopped from going upstream by waterfalls, but once they get above them, there's nothing keeping them from spilling back over and moving downstream. By doing so, introduced guppies can both occupy previously guppy-free pools and affect naturally occurring guppy populations by introducing new genetic material.

Introductions also have a scientific cost. Put guppies into a pool and another scientist cannot use that pool to study what happens in the absence of guppies. As their genes drift downstream, the altered genetic landscape affects potential research throughout the stream's drainage.

I asked Reznick about these criticisms. His response is that guppy introductions are not at all equivalent to moving a non-native species from one part of the world to another. Rather, he has mimicked what goes on in nature by moving guppies from lower parts of a river to upstream locations in the same river drainage. Guppies do, indeed, manage to sometimes get above waterfalls by wriggling up temporary streams flowing down the hill during the rainy season. And he's seen flash floods wash guppies out of streams. So, the presence or absence of guppies in a particular locality is not constant—they come and they go. Indeed, genetic studies in Reznick's lab have shown that guppy populations high up in rivers today have been there a relatively short period of time, the result of recent colonizations. In other words, what we see today reflects the natural balance of colonization and extinction. Just because an upstream pool doesn't contain guppies today doesn't mean that it never has and never will. Guppy populations colonize upstream pools all the time—Reznick's introductions are just an imitation of a natural, ongoing process.

At its heart, this is a philosophical difference about scientific progress and the sanctity of nature in a changing world. There is no objective right or wrong, just a difference in opinion. In Trinidad, such introductions are not banned, and they continue with official approval.

———

EXPERIMENTAL EVOLUTION STUDIES on guppies are continuing, expanding even, as researchers investigate new aspects of guppy evolution, as well as how other species are evolving in response to the guppies. More experimental introductions are being initiated, by Reznick and others. Nonetheless, one central message of these studies is already clear: guppies evolve in predictable ways in response to new selective conditions.

Endler's experimental introduction study, published in 1980, quickly became a classic, followed just a few years later by Reznick's report on life history evolution in guppies from a second introduction. The scientific world took note—evolutionary biology could be an experimental science, even in natural settings. Yet, surprisingly, many years elapsed before the next experimental evolution study was published, and that research had been set up to study something completely different.

chapter six

Lizard Castaways

I f you visit the Bahamas, regardless of where you go, there's one thing you will definitely see. Not beaches, not casinos, not palm trees. Well, probably palm trees—it would be hard to go to the Bahamas and not see those. But even more ubiquitous—on trees, sidewalks, buildings; in the bushes, on the ground, pretty much everywhere—are small brown lizards, members of the genus *Anolis*, the lizards I study.

The lackluster common name for this species, the brown anole, doesn't do justice to this lizard's magnificence. Admittedly, at first glance, these six-inch-long lizards are a rather humdrum shade of brown, though sometimes with elegant diamond or V-shaped markings in black and white running down the back. But all of a sudden, a male will tilt his head upward, often at the same time straightening his forelegs to stand taller, and from beneath his neck his brilliant red-orange dewlap will appear. The ostentatious necktie matches his jaunty attitude—running, posturing, fighting, eating, getting into a little hanky-panky—there's rarely a dull moment in brown anole land.

It's hard to find a place in the Bahamas lacking these lizards, but

A pair of brown anoles

Tom Schoener succeeded in doing so. Now one of the most prominent ecologists in the world, Schoener cut his teeth studying Caribbean anoles, figuring out how so many species can manage to coexist in a single place. In the mid-1970s, Schoener and his biologist wife, Amy Schoener, spent two summers sailing through the Bahamas, surveying islands large and small. The Bahamas is usually described as an archipelago of 700 islands, but that number is a huge underestimate, the result of a definitional sleight of hand—the tiniest islands, craggy limestone with scraggly bushes and, sometimes, small trees, are officially termed "rocks." Thousands of such rocks are scattered throughout the Bahamas, and the Schoeners visited many of them. They found that as rocks decreased in size, they became less and less vegetated; the tiniest, just a few tens of square feet in size, had meager scraps of vegetation.

And, indeed, unlike everywhere else in the Bahamas, such tiny islands lacked lizards.

The Schoeners decided to do an experiment. Clearly, anoles could not survive on small islands. But why not? Even today, the way that populations go extinct is not well understood. The Schoeners saw these islands as an opportunity to examine the extinction process by placing small numbers of lizards on these islands and watching the populations dwindle away to nothing.

It didn't work out that way. The Schoeners monitored the islands for five years. The populations on the tiniest islands, specks barely larger than a hot tub, disappeared quickly. Populations on slightly larger islands lasted a bit longer, some hanging on for four years before succumbing. But on all islands with vegetated area bigger than a pitcher's mound, the lizard populations survived, even thrived. This result was unexpected. If the islands were suitable for lizards, why were they absent? The Schoeners suggested that perhaps periodic catastrophic events were responsible. And when you talk about catastrophes in the Caribbean, one phenomenon is at the top of the list: hurricanes.

THE SCHOENERS' PAPER was published in 1983, but I didn't read it until several years later, when I was finishing my doctoral research on the replicated adaptive radiation of *Anolis*. Oblivious to the ominous foreshadowing of hurricanes, I saw in this paper not a story about survival and extinction, but an inadvertent experiment on evolutionary adaptation.

My doctoral work had documented how anole species adapted to using different habitats. One aspect of this adaptation involved the length of their limbs—species using broad surfaces evolved longer legs, those using narrow surfaces, shorter legs. Given that the islands to which the Schoeners had introduced lizards varied in vegetation, their study constituted an experimental test of the patterns produced by millions of years of evolutionary adaptation: if short-term and long-term evolution occur in the same way, then anole populations on islands with only scraggly, narrow vegetation should evolve shorter legs, whereas those placed on islands with broader vegetation, more like the habitat on the big island from which they had been taken, should maintain their longer legs.

I had wanted to conduct evolution experiments ever since hearing John Endler speak on his guppy research years before. This was my chance. All I had to do was convince Tom Schoener that it was a good idea.

My opportunity came several months later, at a national conference. I contacted Schoener beforehand and we arranged to meet during a coffee break. Nervously, I made my pitch, pointing out that by introducing lizards onto islands with different vegetational characteristics, he had essentially set up an evolution experiment testing the effect of environmental conditions on lizard adaptation. To which he responded in the way I could only have hoped for in my wildest dreams, by inviting me to join his lab at the University of California, Davis, to investigate whether the populations had evolved as I predicted. Two years later, in the spring of 1991, I found myself on the small island of Staniel Cay in the middle of the Bahamas.

WHEN I TELL PEOPLE that I do my field research in the Bahamas, they try to suppress it, but I can see the beginning of a sly grin at the corner of their mouths. I know what they're picturing: beaches, palm trees, hammocks, mai tais in glasses with little umbrellas poking above the rim.

Staniel Cay wasn't like that at all. For one thing, the drink of choice is a bahama mama, sans umbrella. More significantly, there wasn't much in the way of beaches, and most of the vegetation is scrawny dry forest, with only a scattering of palm trees. Instead of fabulous resorts and luxurious, secluded villas, there was just the Staniel Cay Yacht Club, ritzy only in name. Where the club did excel was in its macaroni and cheese dinner special; also, its enormous flying cockroaches were second to none. The rooms were shabby, and the clientele was a motley band of bankers, sailing bums, and small-time drug runners.

My job was to capture as many lizards as possible from the fourteen island populations established by the Schoeners more than a decade before. The goal was to see whether the populations—all founded from the same source—had diverged in leg length to adapt to the vegetational differences among the islands.

Catching brown anoles is one of the more enjoyable aspects of fieldwork and can be accomplished in many ways. The easiest is to go out at night and snatch them while they're snoozing. Anoles sleep in very exposed conditions, on leaves and at the ends of thin branches. Such bed sites allow them to dream in peace, knowing that vibrations from any approaching nocturnal predator will rouse them from their slumber in time for a getaway. This strategy evolved to thwart the likes of snakes, rats, and centipedes, which all have to traverse a branch to get to the lizard, but it's woefully inadequate against a two-legged predator with a flashlight. In the light's beam, the lizards show up readily against the background greenery. Some perch high in the trees, but most are within reach and the only challenge is grabbing them—clapping them gently between two hands is my technique—before they're awakened by the light.

The second lizard-roundup method is more sporting, involving lassoing active lizards. To do so, I fashion a small noose out of some sort of stringy material—my preferred twine is waxed dental floss, white, not minty green, because the latter can't be seen amidst the vegetation. Attaching the lariat to a ten-foot fishing pole, I'm ready to go lizard hunting. Once I spot a lizard—in the best case, perched on a tree in its surveying posture, pointing down and head raised slightly off the vertical surface—I approach slowly until I'm about a dozen feet away. Some lizards won't let you get that close, but brown anoles often will. Then, even more slowly, I maneuver the noose on the end of the pole toward the lizard and slip it over the lizard's head. Why would a lizard allow this to happen? To them, this slender piece of white string is unfamiliar, but not threatening—indeed, sometimes they try to catch and eat it.

If all goes well, I get the noose around the neck and then quickly pull backward. The weight of the lizard's body as it dangles from the pole causes the noose to pull close. Lizards have strong necks and they don't weigh very much, so this does nothing more than surprise them, hurting only their dignity. They're certainly not happy about it, though, as their now-open mouths attest, and they miss no opportunity to bite me as I remove them from the noose. Lizards do have teeth, sometimes quite sharp, but at the brown anole's size, they rarely break the skin.

If only lizard hunting were so simple! Alas, this description makes the process sound easier than it is. These islands are not the easiest place to work. Made of porous limestone, the rock on the islands is heavily eroded, leading to holes, jagged edges, and pieces that unpre-

dictably break underfoot. The vegetation is ragged, in some places thick and in others dominated by poisonwood, a nasty relative of poison ivy capable of growing into a tree.

And the lizards themselves can be difficult as well. Although some will stand stock-still and allow you to slip the noose over their necks, most are at least somewhat wary, moving their head slightly away as the noose approaches, even squirreling around to the backside of a branch. Plus, vegetation impedes positioning of the noose, and ill-timed gusts of wind blow it off track. For these reasons, I liken lizard catching to fly-fishing, a primordial battle between a person with primitive equipment versus an animal with a miniscule brain. I've explained this idea to many fly-fishing enthusiasts and usually have been met with looks of disbelief and incredulity. Apparently, fly-fishing is some elysian experience with no equal on this earthly plane. Nonetheless, take it from me, the challenge of lizard catching is every bit as much a battle of man versus nature, one in which I am often outwitted by the pea-brained saurian.

There's yet one other way to nab your anole. Walk right up to one and grab it with your hands. This trick never works for me—they sense my intent well before I'm close enough and skedaddle away. But one of my colleagues, Manuel Leal—born and raised in Puerto Rico around anoles—is able to sidle up next to them and then, with a lightning-quick thrust of his hand, pluck the lizard off the tree. I still haven't figured out the magic of this lizard whisperer.

For this particular study, the actual act of capturing the lizards was only part of the challenge. The bigger issue was getting to the tiny experimental islands in the first place. From Staniel Cay, the islands weren't very far away, all within several miles. But growing up in Saint Louis, I didn't learn much about boating, particularly about fixing balky boat engines. Every day as I went out in the Boston Whaler I rented from the club, the question was whether I would make it back or whether I'd have to wait for the staff to realize I was missing and send someone out to find me. I took to bringing along a book to read while awaiting rescue.

The scariest moment of the trip was the day when the Whaler's motor conked out right next to a large island—much bigger than the experimental rocks—that bore a house, a warehouse-like building, and an airplane strip. The island was owned by a man reputedly engaged in shady activities (this was during a time when the drug trade was running rampant in the Bahamas, as a conduit from South America to the United States). I was warned to avoid that island and its unsavory inhabitants at all costs. And there I was, drifting right offshore with a dead engine. With great trepidation—I'm not sure I've ever been more scared in my life—I swam the boat ashore, walked up to the door, and knocked. A friendly man answered, I explained my predicament, he radioed the club, and fifteen minutes later I was back in business. Who knew drug traffickers were so nice? Or maybe they just appreciated a good experiment?

Once I captured the lizards, I placed them in little bags and put them in a cooler to keep them from overheating. At the end of the day, I returned to my bug-infested room (low point of the trip: when I poured my Raisin Bran cereal into a bowl for breakfast and out spilled half a dozen roaches). There, I put the lizards to sleep with veterinary anesthetic and quickly measured the length of their legs before they revived. The next day they were returned to their islands and released at the exact point of capture, unharmed and with a great story to tell their mates.

The work went much more slowly than anticipated. It was a dry, windy spring, terrible conditions for lizard catching. The lack of rain suppressed insect activity, meaning there wasn't much food around for the lizards, and the combination of sun and wind promoted dehydration. Sensible beings that lizards are, they hunkered down out of sight, waiting for better times. Midday was the worst—all lizard activity ceased. I went through all of my reading material. But when the trip came to an end after four weeks, I had succeeded in capturing and measuring 161 lizards.

This was in the days before laptops. Every time I took a measurement, I wrote it down on a piece of paper. Curious about what the data were showing, I went old school and plotted the data on hand-drawn graph paper. My graphs didn't show any obvious patterns. This agreed with the sense I had from handling the lizards—they didn't obviously differ from one island to the next. I can't say that I was surprised—the populations were young. Perhaps it was too much to expect that they might have evolved in such a short time span.

I returned to my office in Davis and got busy with other ongoing projects. The data weren't forgotten, but they just weren't a high priority. I already knew there was nothing there, so was in no hurry to plug the data into my computer to confirm the non-news. Eventually, I crossed everything else off my to-do list and set about inputting the data into my computer's statistics program. At last, the time came to conduct the formal analysis.

At first, I misread the results on the screen, interpreting them as confirming my expectation that nothing notable had happened on the islands. Then I looked again and the realization hit me. The populations not only had evolved, but had done so exactly as we'd predicted: on islands where lizards used thin branches, they tended to have very short legs, whereas on islands where they used broader perches, their legs generally were longer. We'd experimentally demonstrated rapid, adaptive evolution in nature (needless to say, that was the last time I used homemade graph paper to look at preliminary data. In my defense, the differences in leg length among populations, though statistically significant, were small and thus not obvious on a hand-drawn graph).

It took us some time to finish all the analyses and write the paper.

By the time it was about to be published in the British journal *Nature*, Tom Schoener and I were back in the Bahamas for another field trip (along with colleague David Spiller), this time working on a new experiment on the northern Bahamian island of Abaco. The accommodations there were definitively an upgrade—better rooms, better food, fewer cockroaches. But still no phones or internet in our rooms.

Not knowing what was to come, I had changed my office answering machine message before I left, saying that I could be contacted by leaving a message with the front office of the tiny motel where I'd be staying. Unbeknownst to me, the PR people at *Nature* had put out a press release that said, "This may be among the most important work in evolutionary studies since Darwin studied the diversity of finches on the Galapagos Islands during the voyage of the *Beagle*." Now, it was a nice paper, but that was way over the top.

Midway through the trip, I returned after a long day out on the islands to find a message from the hotel owner, who doubled as the onsite manager. I walked over to his office, and he informed me that a reporter from the *New York Times* had called. The next day it was the *Boston Globe* and *USA Today*. The day after, ABC News wanted to arrange to send a team to report from the Bahamas.

The owner was clearly taken aback. After many years as a hotelier, he thought he'd seen it all. It was odd enough for someone to come all the way to the Bahamas just to run around chasing lizards. I seemed harmless enough, but clearly I had a screw loose. And then all of a sudden, the world's media was beating down his door trying to get hold of me and, incidentally, tying up his sole phone line. Maybe it was a coincidence, but shortly thereafter he put the place up for sale.

We returned from the Bahamas just as the story broke (*Nature* issues its press releases a week before publication, but insists on an embargo preventing announcement of the results until the day the paper is officially published). I had my fifteen minutes of fame. I have to admit that it was a thrill to see my name in the first section of the *New York Times* and on the front page of the *Boston Globe*, not to mention in *USA Today* and many other newspapers and magazines. ABC News aired a report, though without onsite reporting. I received congratulations from friends and colleagues near and far. The story line was that we had demonstrated remarkably rapid evolution, and we had done so by conducting an experiment in nature. Endler's and Reznick's work notwithstanding, this was still big news.

As with the guppy work, our study investigated the predictability of evolution by testing observations from natural variation. Adapting over millions of years, anole species vary in limb length depending on the diameter of surfaces they use. Would the same result evolve over the course of a few years if we placed initially similar populations on islands differing in vegetation? The answer was yes. After a decade of evolution, our fourteen populations differed in limb dimensions, the length of their legs proportional to the width of the branches they used. Like the guppy studies, we could predict how lizards would evolve— re-create conditions experienced by natural populations and the experimental populations would adapt repeatedly in the same way.

BUT JUST AS WITH THE GUPPIES, we had to consider one other possibility, that the differences in leg length among the populations were not the result of evolved genetic changes. When I gave scientific talks on the work, there would always be someone in the audience—usually a pesky botanist—who would raise the question of phenotypic plasticity. Were the differences among the populations actually the result of genetic change? Could it be that lizards born on islands with narrower vegetation simply grew shorter legs?

It seemed to me implausible that a lizard's leg length could be affected by the diameter of the surfaces it used. How would using a narrow perch as a baby lizard cause its legs to grow less? But I kept getting the question, so I knew I had to look into it.

I hit the library to find out what was known about the effect of perch diameter on limb growth in lizards. That didn't take very long, because no one had investigated the topic. There was a broader literature of relevance, however, on the effect of exercise on limb growth in vertebrate animals. The overarching goal of this work was to determine whether exercise of various sorts affected how an animal's legs developed as it grew. These studies included some of the most bizarre experiments I have ever read.

For example, in one study the investigators forced young lab mice to run on an exercise wheel for ten hours a day, while the control group just lazed around in their cages. In another, young rats were tossed in a bathtub for four hours a day and made to swim; again, doldrums for the controls. Or, in yet a third, growing chickens were put on a treadmill and made to run for long periods.

The results of these experiments were fairly consistent. Animals induced to exercise for long periods grew thicker limb bones. There's even a well-known parallel in humans: weight lifters have thicker arm bones than other people. The reason is that bone is actually a very dynamic substance, adding and losing calcium all the time. When stress is placed on a bone, as during exercise, the bone adds calcium to strengthen itself. As a result, the width of a bone is a plastic trait affected by the behavior of the animal.

But our results were not about the *width* of the bone. We were studying its length. And for the most part, these studies had not found differences in limb length resulting from exercise. There was, however, one major exception, an old study from the 1950s on male professional tennis players. If you think about it, pro tennis players have been swatting balls since they were little, placing consistent and great stress on their serving arms throughout their growth years. And the charm of this study was that each individual could serve as his own control, by comparison of the serving and non-serving arms.

And, indeed, their serving arms were longer; smacking a tennis ball for years on end does, indeed, make your arm grow longer.* Measurements were made on x-ray images, so the differences were definitely in the length of the bone and had nothing to do with ligaments or muscles.

Apparently, using limbs in different ways during growth can affect how long a limb grows; the hypothesis of phenotypic plasticity wasn't completely far-fetched. But it's a long way from professional tennis players smashing forehands to lizards hanging on to branches. It was clear that we had to do a plasticity study.

The most straightforward approach would have been to capture baby lizards (or mothers with eggs) from our study islands and grow them all in a common lab environment to see if the differences persisted. However, our studies were ongoing and the populations were small. We were afraid that if we removed a lot of lizards from the islands, we would affect the future of the experiment. So a common garden experiment was out.

We went to plan B, the opposite of a common garden study. Instead of taking lizards from different populations and growing them in a single location, we took lizards from one population and grew them under different conditions. One group of lizards was raised in terraria with a broad flat piece of wood (a two-by-four) upon which they sat. In contrast, for the other group, the only perch was a very narrow dowel a quarter of an inch wide. The experiment tested whether growing up using such different surfaces would affect how long a lizard's legs grew to be. In other words, could phenotypic plasticity produce differences comparable to those we observed among our populations in the field?

---

* Of course, in theory the causality could run the other way. Maybe only people with asymmetric arms become professional tennis players.

I conducted this experiment just to shut up the annoying botanists, to demonstrate that just because plants grow differently in different conditions, that doesn't mean that lizards' legs do the same. I was quite surprised, then, to look at the data and see that I was wrong. Damn botanists! Put a growing lizard on a broad surface, and its legs end up longer than lizards raised on narrow rods, even after accounting for differences in overall body size.

Nonetheless, the study also suggested that the differences observed among our experimental islands were too great to be explained by plasticity. The lab growth experiment subjected the lizards to much greater differences—a narrow rod versus a broad piece of wood—than the differences in perch diameter among the experimental islands. Yet the difference in leg length among the island populations was three times greater than the difference produced in the lab. In other words, even under extremely different conditions, phenotypic plasticity could only account for a fraction of the variation seen on the islands. Consequently, we concluded that evolved genetic changes are likely responsible for the majority of the differences in limb length seen among the experimental island populations.

Of course, this is an indirect way to test for a genetic basis for limb length differences. Back when this work was done, it was not possible to directly examine the genome and locate the genes responsible for determining leg length. Twenty years later, we're still not quite there, but within the next few years, researchers likely will have identified the relevant genes and we will be able to discover which genetic differences are responsible for variation in limb length among populations.

IMAGINE YOU'RE AN ANOLE living on a small, isolated island in the Bahamas. You spend a lot of time on the ground, catching insects and consorting with your colleagues. Then one day, out of nowhere, a couple of big, oafish lizards show up on the island. They're kind of clumsy and can't climb worth a darn, but they've got big mouths and make clear that they'd love to have you for dinner. What do you do?

The answer is obvious to anyone with a lick of sense. Climb up into the bushes, stay off the ground, and keep away from the brutes. But then you've got another problem—your legs are too long to easily navigate the narrow branches. You've got some evolving to do.

And that, in a nutshell, was what our next experiment was about. After the success of the first study, Schoener, Spiller, and I decided to try another experiment, this one intentionally set up to look for evolutionary change. Again, we were testing a prediction based on observations from nature. Schoener's travels through the Bahamas two decades previously had revealed that brown anoles perch higher in the vegetation on islands inhabited by a larger and more terrestrial lizard species. Our prediction this time was twofold: first, that in the presence of the predator, brown anoles would move upward off the ground and into the bushes, and, second, that they would subsequently adapt to this new habitat-use regimen, again evolving shorter legs to move about on the narrower surfaces.

The general framework was similar to the previous study, focusing on brown anole populations on tiny limestone islands. But this time, we used slightly larger islands on which the anoles were already present and instead introduced the larger, ground-dwelling lizard eater.

The bad guy in our experiment was the curly-tailed lizard, a stocky lacertilian that grows to twice the length and ten times the weight of brown anoles. The curly-tailed lizard's name is self-explanatory—when bothered, the reptile runs away with its tail rolled up in a curlicue above its body. Why it does this is not known. Perhaps it's sending a message to the predator—"I see you. Don't waste your time chasing me"—or

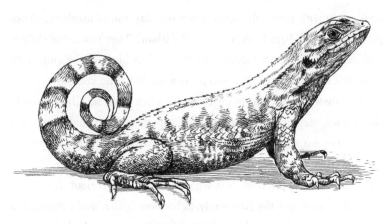

Curly-tailed lizard

perhaps it's trying to divert attack to its expendable tail. Regardless, it's a comical site to see an ungainly lizard trundling away with its tail coiled above its body.

Anoles, no doubt, are less amused by the lizard clown's presence. Curly-tails will eat anything they can get into their cavernous mouths, and that includes other lizards.

Although our experiment sounded good in concept, we were not at all confident of the outcome. Curly-tailed lizards had been reported to eat anoles, but we had no idea how important an ecological factor such predation was. Did it occur just once in a blue moon with no overall effect, or would it have a major impact on the anoles? No data were available to answer the question. We'd have to do the experiment and find out.

Curly-tails occur on the larger rocks around Abaco and occasionally colonize nearby smaller ones, so our translocations were mimicking an ongoing natural process. At the outset, we located twelve islands and sorted them by size and vegetation cover into six pairs of islands.

Then, by the flip of a coin, we decided which island in each pair would receive five curly-tailed lizards and which would serve as the control.

The curly-tail roundup took place in April 1997. Capturing curly-tails is even more fun than capturing anoles because we have to use a longer, twenty-foot pole, the reason being that curly-tails are warier than anoles and won't allow us to approach as closely. As a result, better motor control is required to maneuver the pole over the lizard's head, particularly on a breezy day, making curly capturing a greater challenge. The end result, though, is the same—a lizard dangling from a pole, dental floss lasso around its neck. The only difference is that greater care must be taken in removing curly tails from the noose because they have much larger mouths than anoles and thus their bites actually hurt.

We returned three months later for the first follow-up, not knowing what to expect. Would the curly-tails have survived on their new homes? Would their presence have made any difference to the anoles? We had our predictions, of course, but not a lot of confidence in them.

To our surprise, the results were already striking. The size of the anole populations on the curly-tail islands was half that on the controls, a difference that was maintained for the rest of the experiment. On the control islands, the anoles continued to be found on or near the ground, but on the curly-tail-introduction islands, they had shifted upward, off the ground and away from their nemeses. Two years into the experiment, anoles on curly-tail islands were perching on average seven times higher than on the curly-tail-less control lizards.

These results were more dramatic than we had anticipated. The anoles had moved up in the bushes and were using narrow vegetation. Just by watching the lizards move clumsily on these narrow surfaces, we could tell that they were not well adapted. Our prediction was that natural selection would work its wonders, and that in a few years the lizards would evolve shorter legs and become better adapted to moving in their new arboreal habitat.

Alas, we never got a chance to find out. In September 1999, Hurricane Floyd, a Category 4 monster, scored a direct hit on Abaco. Our experimental islands, just several feet above sea level, were submerged under the storm surge for several hours. All the lizards were washed away. The experiment was over.

This was actually the second time in three years that we'd had an experiment ended by a hurricane. We had moved our operations to Abaco after Hurricane Lili went right over our heads in Georgetown, Bahamas, in October 1996, wiping lizards off another set of islands. We learned a lot about the impact of hurricanes from these events, including confirmation of the Schoeners' insight about why small islands are lizardless. Although we'd accidentally become hurricane experts, it had come at the cost of premature termination of several of our long-term experiments, a costly trade.

But there was a silver lining. Now that we knew that curly-tailed lizards have such a major impact on anoles, we could take advantage of that knowledge to redesign our next experiment. And another stroke of fortune: Floyd had occurred earlier in the season than the previous hurricanes, before the lizard breeding season had ended. Even though all the lizards had been wiped off the islands by the storm's surge, their eggs in the ground had remained in place. A month later, to our amazement, the island was covered with baby lizards, hatched from eggs that had survived a six-hour submersion during Floyd's high waters.

We still had to wait a few years for the vegetation and lizard populations to recover, but by 2003, we were back in business: Curly-Tail Introduction II, the sequel. The plan was generally the same: introduce curly-tailed lizards to some islands and not others. But this time, we did something different. Our goal was not only to track the populations through time to see if they evolved, but also to actually measure natural selection itself.

Specifically, our hypothesis was that the presence of curly-tailed lizards would alter patterns of natural selection. Our prediction had two parts. Initially, we expected that longer-legged anoles—faster and thus better able to elude curly-tails on the ground—would survive better. But over time, we expected the anoles would shift their habitat use from the ground up into the bushes, as they had in the previous iteration of the experiment. Once off the ground and away from the terrestrial curly-tails (which are too large to climb all but the broadest trees), long legs would no longer be an advantage. Rather, as in our Staniel Cay experiment, we expected natural selection to favor shorter-legged lizards, the ones more adept at moving on narrow surfaces.

Natural selection rewards those individuals that produce the most offspring that survive to the next generation. There are a number of ways to maximize this reproductive success: by surviving to an old age, by maximizing the number of mating episodes (referred to as "sexual selection"), and by maximizing the number of offspring per reproductive event. In this case, we were examining how well adapted the lizards were to their environment, so we chose to examine survival as our metric of evolutionary fitness.

To determine whether survival was related to leg length, at the outset of the experiment we needed to capture anoles, measure them, and give them a unique identifier so we could determine how long they survived. Would lizards with shorter legs live longer? To find out, before introducing the predators, we visited all of the islands and captured as many anoles as possible.

When ornithologists want to identify individual birds in a population, they put little plastic-colored bands on their legs. Each bird gets a unique combination of colors on the two legs, allowing the scientist to identify each bird from a distance with binoculars (right leg: orange on top, then two black bands; left leg: yellow, orange, yellow—it's Fred!). Anoles, however, are too small for such bands, or for the microchips that vets inject into cats and dogs. Marks on their skin would be lost every time a lizard sheds, which occurs frequently in the summer. So herpetologists use a method developed for marking salmon, injecting colorful non-toxic rubber threads just below the skin. Because the lizard's skin is translucent on the underside of its legs, the neon color of the elastomers—Day-Glo green, yellow, pink, orange—is readily apparent when the lizard is recaptured. By varying colors and which parts of the limb were injected, each anole received a unique color code.

Learning from our previous studies, I realized that it was more efficient to process the lizards on the islands immediately after capturing them, rather than transporting them back to the room, keeping them overnight, and then returning them to their homes the next day. To do so required setting up a mobile workstation. For the most part, the islands are composed of prickly limestone rock, unsuitable as a working or sitting surface. So I borrowed a plastic outdoor chair from the room I was renting and took it with me on the motorboat to the islands.

In some respects, this was a sweet location for a lab. The islands were quite small, so I was always a few feet from the ocean. Sightings of rays and turtles were common, and occasionally a pod of dolphins would swim by.

On the other hand, there were no trees to sit under—the islands were completely exposed to the blazing Bahamian sun. When there was no breeze, the midday heat was stifling, exacerbated by my head-to-toe sun-protection apparel. At least my enormous sun hat, the size of a small flying saucer, provided some shade, even if it made me the laughingstock of the boats full of tourists that passed by. Windy weather was a mixed blessing. It definitely kept me cooler, but came at a cost, as my materials and hat were at risk of blowing away.

My procedure was to catch a lizard, walk back to my chair, sit down, and measure the lizard, being careful not to drop the squirming saurian as I recorded the data in my notebook. Then, I would reach over to the cooler containing the four syringes, each loaded with a different color and chilling on ice to prevent the liquid from hardening prematurely. The syringe would be inserted just under the skin and the color squirted in. The liquid quickly hardened into rubbery solidness and the lizards were then returned to their exact point of capture. The whole process was over in ten minutes.

It took about a month, but eventually we caught almost all the lizards on the twelve islands. Each lizard was now uniquely marked—we could go back, catch that lizard again, and know its identity. And, most importantly, we knew all about that lizard: how big it was, how long its legs were, how many scales it had on its toepads. We were primed to see if survival was related to phenotype: would lizards with shorter legs survive better than those with longer legs? And, critically for our experiment, would the presence of curly-tailed lizards alter the operation of natural selection?

Once all the lizards were measured, we held another curly-tail roundup on Abaco and introduced the lucky winners to their new island resort homes. Again, six islands received the predators and another six served as controls. Then we flew back home, leaving it to the lizards to sort things out.

Six months later, over the Thanksgiving holidays, we returned to see what had transpired. Our goal was to capture every single anole on every island to determine who had survived and who hadn't. Doing so is not easy. Catching the first eighty to ninety percent isn't hard, but getting the last few is tricky—always a couple continue to elude capture, poking their heads out, ducking under cover, quietly remaining out of sight.

Once captured, lizards were turned onto their backs for us to inspect the undersides of their legs. The colors were usually easily detected, but just in case, we brought along a flashlight that emits ultraviolet light because the threads glow under UV. Once the examination was complete—which took about a minute—a tiny dot was placed on the lizard's back so that we'd know it had already been caught and then it was let go at its place of capture.

Our hypothesis was that natural selection had operated on leg length. To test this idea, we calculated what is called a selection gradient, which in this case was basically the difference between the leg length of the survivors versus that of the lizards that perished. A large positive value would indicate that long-legged lizards survived better, a large negative value would indicate the reverse.

On the control islands, sans curly-tailed lizards, selection gradients were about zero—limb length was not affecting survival. But on the curly-plus islands, the story was different. Selection gradients were all very high and positive. Long-legged lizards were surviving better in the presence of curly-tailed lizards; the presence of predators was altering natural selection in exactly the way we had predicted.

While we were capturing lizards in November, we also recorded where we found them. Just as we saw in the previous experiment, the anoles were getting into the bushes to avoid the curly tails—lizards were on the ground one-third of the time on control islands, but only in ten percent of observations on the islands with curly-tails. Moreover, on the curly-tail islands, anoles perched higher and on narrower branches.

As a result of this change in habitat use, we expected that natural selection would eventually change direction. Out of the reach of curly-tailed lizards, long legs would no longer be advantageous. And we know how anoles adapt to using narrow surfaces—they evolve shorter legs for better maneuverability. So we expected that natural selection would eventually start favoring shorter-legged lizards on the predator islands.

We returned to the islands the following May for our next census. Again, we captured all survivors. We noted that the habitat differences were even greater, as the lizards on the curly-tail islands were spending even less time on the ground and using narrower perches. We again calculated selection gradients, this time considering only those lizards present the previous November, and comparing those that had made it to May to those that had died in the preceding six months.

Again, selection gradients on the control islands were around zero—natural selection continued to ignore limb length on these islands. But on the curly-tail islands, the story had changed. Natural selection was again operating, but this time in the opposite direction. Shorter-legged lizards were now surviving better—natural selection had completely reversed. We had expected this to happen, but not nearly so quickly.

These results documented natural selection within a generation, but not evolutionary change across generations. In fact, the two selection episodes on the curly-tail islands pretty much canceled themselves out, so that the net selection was around zero. But we didn't expect natural selection to continue to oscillate from positive to negative in the future. Now that the anoles were up in the bushes, they weren't coming back down. The curly-tailed lizards would see to that. Moving forward, we predicted that natural selection would continuously favor shorter legs. We looked forward to seeing whether brown anoles would evolve to become similar to the twig specialists on the Greater Antilles.

Based on our previous experiments, you might be expecting me to tell you that our experiment was wiped out by a hurricane. But if that's what you're thinking, you're mistaken. A hurricane did not terminate this experiment. Two hurricanes did. Hurricanes Frances and Jeanne delivered a one-two punch three weeks apart in September 2004 that finished this experiment before any evolution could occur.

As with the previous experiment, the curly-tailed lizard populations were eliminated, but most of the anole populations survived, albeit in greatly diminished numbers. The vegetation on the islands was pummeled. We had to wait four years, but we started the experiment again in 2008. As I write this, the experiment is ongoing—knock on wood—though it hasn't been easy thanks to several more hurricanes. We hope to have results soon.

Just as before, however, there was a silver lining to the 2004 hurricanes. While waiting for the bigger islands to recover, we decided to initiate a new experiment on some even smaller islands, about the size of a large living room, that had been cleared of lizards by the hurricanes.* This experiment took a slightly different form. We collected lizards from a large, heavily forested nearby island and introduced them to seven islands with particularly scrawny vegetation. In other words, the populations went from living on tree trunks and broad branches to narrow stems and twigs. Our prediction was that they would evolve shorter legs.

And they did. Over the course of four years, average limb length steadily declined on all seven islands. The lizards were evolving exactly as predicted, and the extent of change was substantially greater than produced in our lab plasticity experiment. This study was going particularly well, turning into an especially nicely documented example of rapid evolutionary change. The populations even survived Hurricane Irene in 2011. Alas, Hurricane Sandy the following year was a different story, wiping five populations off the map. We're continuing to monitor the two populations that survived, but the results were much more compelling when all seven islands were evolving in lockstep.

Frankly, I'm getting a little tired of hurricanes.

---

* Schoener, Spiller, and I were joined on this project by Jason Kolbe and Manuel Leal.

DESPITE THE GREAT ATTENTION paid to the guppy and lizard research, few other researchers followed our lead. One impediment, no doubt, was the time and effort needed to bring such work to fruition, not to mention the possibility that freak meteorological—or other—events could intercede and destroy the project after years of work. Moreover, although our studies had shown that detectable results could emerge after only a handful of years, there was no guarantee that other organisms would evolve that quickly. What if several decades, rather than several years, were required before evolution became evident?

But there is another way to experimentally study evolution, one that paradoxically doesn't require putting in years of work, yet allows researchers to study the results of decades of evolution. Although long-term evolution experiments were a novel idea in the 1970s and 1980s, long-term ecological studies were not.* Indeed, our first study, on Staniel Cay, had hijacked an experiment set up to test an ecological phenomenon: was lizard population survival related to island size? Unintentionally, the Schoeners' experiment had set the stage for me to go back later and see if evolution had occurred on the experimental islands. As a result, I was able to look at the result of ten years of evolution without having to set up the experiment and wait a decade.

It turns out that our study wasn't the only one that could be retrofitted to study evolution. One ecological study in particular had been running for well over a century, making it the reigning granddaddy of evolution experiments.

---

* To scientists, "ecology" is the study of how organisms interact with their environment. The term was expropriated in the 1970s by the environmental movement to take on its broader meaning, more or less synonymous with "natural environment."

chapter seven

From Manure to Modern Science

More than 170 years ago, the longest continuously running experiment in the history of science began on fields thirty miles northwest of London. From his boyhood, John Bennet Lawes had been fascinated by how plants grow. During his time at Oxford, he became interested in growing medicinal plants on the family estate, Rothamsted, but his attention soon turned to developing means of enhancing agricultural productivity. In turn, this led to experimentation on "artificial manures," and by age thirty, he had founded a company that led to the rise of the chemical fertilizer industry.

In 1843, Lawes decided to turn his manor into an agricultural research station (long known as the Rothamsted Experimental Station, but recently rebranded as Rothamsted Research). He hired a chemist, Joseph Henry Gilbert, and together they concocted a plan to use Rothamsted's fields as a test site for experimentation on the effects of various fertilizers on crop growth. They began many experiments over the next decade and a half, and seven have been continuously maintained ever since. These experiments investigate the efficacy of different fertilizers, crop rotation, and harvesting schedules on plants such as wheat, barley, turnips, and potatoes.

The importance of these experiments for the development of modern agriculture was immense. On Lawes' death in 1900, the *Times* of London remarked:

> To indicate ever so briefly the scope of the investigations which have been successfully conducted at Rothamsted would be, in effect, to summarize the history of the progress of agricultural chemistry during the last half-century.... Sir John Lawes was one of the greatest benefactors of agriculture—perhaps the greatest— the world has seen. His originality in experimental research and his inflexibility of purpose, coupled with a genius of no ordinary kind, enabled him to discover grand truths which have had a profound influence upon the progress of agriculture.

Lawes and Gilbert started the last of their experiments—now known as the Park Grass Experiment—in 1856 on a seven-acre grassy meadow. Unlike the other experiments, the PGE did not study the factors that maximize the production of a particular crop. Rather, it focused on the yield of harvestable hay. Back in those days, of course, farmers mainly fed hay to their livestock, so high yields of hay were just as important as the production of the crops they took to market.

If you're a city slicker like me, the word "hay" may evoke an image of tied-up bales, perhaps those you sat on during a tractor ride on the weekend your parents took you for a farm stay. But you may not know—as I didn't—that hay is simply whatever plant material is grown in an open agricultural field, cut down, dried, and used to feed livestock. Many different types of grasses, as well as alfalfa and clover, are common hay species.

Unlike the other Rothamsted experiments, the PGE did not involve planting crops every year or several years. Rather, the experiment began with a long and narrow field that had been used for hay production for at least a century. Many different plant species occurred in the field. Lawes and Gilbert divided the field into thirteen strips, each about seventy feet wide, and applied a different mixture of fertilizers to each plot, leaving two plots unfertilized as controls. Periodically (every year to several years), the fertilizer treatment was repeated.

The primary goal of the experiment was to evaluate the efficacy of man-made fertilizers compared to the traditional manures used by farmers. To do so, plots received different fertilization treatments. Most received mixtures of various types of inorganic chemicals (for example, ammonium, magnesium, potassium, and sodium). Other plots received combinations of farmyard manure, pelleted poultry manure, and fishmeal.

Initially, the plots contained a great diversity of different plant species, with little variation among the plots. Critically, and unlike the other Rothamsted experiments, the plots were not reseeded; rather, nature was left to take its course.

The Park Grass Experiment has been maintained continuously now for more than a century and a half. During this time, some aspects of the experiment have been slightly altered. The plots established in 1856 constitute the bulk of the acreage, but over the subsequent sixteen years, another seven were added on the south and west ends of the field, bringing the total number of plots to twenty. A few other changes have been made, the most substantial in 1903, when all plots were divided in two. Both halves continued to receive the same fertilizer treatment they had been treated with since 1856, but in addition, one of each pair of subplots began receiving lime (the calcium-based mineral fertilizer, not the fruit), causing the soil in those subplots to become less acidic.

Very quickly, the Rothamsted experiments confirmed Lawes' and Gilbert's idea that artificial fertilizers were as good as manure in raising hay yields. But the Park Grass Experiment just as swiftly showed something they didn't anticipate. The species composition of the different plots, initially so similar, quickly began to diverge as species disappeared from the plots. This change in species composition was so rapid and dramatic, Lawes and Gilbert wrote, that in less than two years "the experimental ground looked almost as much as if it were devoted to trials with different seeds as with different manures."

These differences are so obvious a century and a half later that they're apparent in satellite images. The different plots appear as little patches side by side, but differing in color: dark green, light green, some almost white, others with a brownish tinge.*

On the ground, the differences are even more obvious. Over the years, most of the experimental treatments have caused a reduction in the number of species on the plots. Provision of ample fertilizers allows the fastest-growing plants to outcompete the others, squeezing many species out of a plot. In addition, some fertilizers cause the soil to become highly acidic, eliminating species that can't grow in such conditions.

Let's take a walk through the PGE. Plot number 3 is a control, left alone for the last century and a half, its soil unsullied by added nutrients. Come for a visit in June, and it's a riot of color and texture. Reds, yellows, greens; flowers and stems in many shapes and sizes. Red fescue, a type of grass, is the star. It dominates the plot, its thin, stiff stalks spouting long-stemmed, reddish-purple flower clusters. But it's supported by a cast of several dozen other species of grasses, as well as herbs, some with large and beautiful flowers.

---

* Check them out yourself on Google Earth by typing in "Rothamsted Estate." Through the miracle of the internet, you'll quickly find yourself hovering above the Hertfordshire countryside. As you magnify the image, you'll find little digital sticky notes just below and to the left of the estate. Put your cursor over one and it will say "Park Grass Experiment." Zoom in a bit more and you'll see the experimental fields.

One of the plots in the Park Grass Experiment

This is typical hay meadow vegetation, and the entire field once had this appearance. But most of the other plots are no longer so richly nuanced. The vegetation on some plots may be thicker and taller, but their composition is much more homogeneous. Nearby, for example, plot number 1 has received an annual dose of nitrogen and other minerals since the experiment began. Now, there aren't all that many species. Several grasses dominate, higher and heftier than number 3's red fescue. Flowers are scarce, poking out here and there.

And then there's plot number 9, which has been treated with ammonium sulfate for the past 150 years. The resulting soil acidity has driven out not only most of the plant species, but also the earthworms and other subterranean soil dwellers. Only three plant species remain; look out over number 9 and you'll mostly see the tufted, hairy tops of sweet vernal grass, a species that occurs on almost all the PGE plots.

The different fertilizer treatments have changed the PGE plots in many ways, altering their soil, affecting how robustly plants can grow, determining which species can coexist. Since the days of Lawes and Gilbert, the differences among the plots have been attributed to these ecological phenomena, whether a species could tolerate the conditions on a plot and whether it could survive with the other species that occurred there.

For more than a century, no one thought to ask whether evolution might be playing a role in creating the differences among the plots, whether populations of the same species on the different plots were adapting to their local conditions. And why would they? Not only was the Darwinian glacial-pace-of-evolution mindset still prevalent, but these plots occur right next to each other, separated in some cases by just a few inches. Standard evolutionary biology wisdom of the time was that genetic exchange between populations—"gene flow"—would prevent divergence. Pollen drifting from one plot to fertilize a plant in another or wind-blown seeds would move genes back and forth, keeping the populations genetically homogeneous.

But a young botanist named Roy Snaydon was not so sure. When he started graduate studies in Wales in the late 1950s, botanists were just beginning to realize that plants could evolve quickly, even in the absence of isolation. His doctoral advisor, Tony Bradshaw, was in the process of publishing his now-classic work on the evolution of heavy-metal tolerance in plants growing on top of old, abandoned metal mining sites. Bradshaw found that on sites where copper, lead, and zinc mines had previously existed (some dating back to the time of the Romans or possibly earlier, in the Bronze Age), the soil was contaminated with high metal concentrations, toxic to most plants. Nonetheless, some species were able to grow there. Bradshaw realized that the plants at the mine site had evolved to live in these toxic conditions after the mines had been established, one of the first clear examples of rapid evolution in nature.

In addition to evolving quickly, the mine-site plants were also able to adapt in the face of gene flow. Only a few feet away from the old slag piles, metal concentrations dropped precipitously. Bradshaw and his students discovered that plants of the same species from the surrounding pristine soil could not grow on contaminated soil. The mine-site plants had evolved heavy-metal tolerance even though they were surrounded by metal-intolerant plants whose pollen and seeds, and the metal-intolerant genes they contained, were constantly being blown into the area. Gene flow apparently wasn't as homogenizing as commonly thought.

Snaydon's doctoral work followed Bradshaw's lead, documenting adaptation to soils with different chemical composition in white clover and a common grass, sheep fescue. After he finished his dissertation, Snaydon became a faculty member at Reading University, where he was introduced to Rothamsted, fifty miles to the northeast. Every year, he took his honors botany students for a visit, and while he did, the wheels began to spin. Snaydon saw in the Park Grass Experiment a way to experimentally test the idea that soil chemistry can drive evolutionary divergence in plants, even over very short distances and short periods of time. If this were the case, he reasoned, then it was possible that the variation seen among the Park Grass Experiment plots may partly have resulted from the adaptive divergence of members of the same species to the varying conditions on the different plots.

There was only one problem: the staff at Rothamsted looked upon the experimental plots—at that point one hundred years old—as hallowed ground. Only a few select staff members were allowed to even walk on the plots to tend them. Nobody was allowed to collect material or conduct research on them. The scientist supervising the plots, Joan Thurston, and the Plots Committee were dubious about Snaydon's proposals, but his request came at the right time. The committee was considering discontinuing the experiments because they saw nothing left to learn, so what could be the harm in letting the professor do a little work on a few plots? Snaydon was called to appear before the committee and intensely grilled. Finally, approval was granted, albeit grudgingly, and they permitted Snaydon to collect a limited number of seeds. Thurston watched with an eagle eye to make sure he didn't exceed his allotment.

To test his idea that plants had diverged among the plots, Snaydon focused on sweet vernal grass, the plant found on the plots throughout

the experimental field. He initially selected three plots that had been fertilized with different chemical mixes since the initiation of the experiment in 1856. Because lime had been applied to the southern half of each plot for half a century, the study involved six subplots varying markedly in mineral content and soil acidity. Snaydon's hypothesis was that over the past century, the grass populations had diverged evolutionarily to adapt to the specific conditions they experienced.

Sweet vernal grass

And diverge they had. Snaydon, quickly joined by ace graduate student Stuart Davies, found tremendous variation in the sweet vernal grass from one subplot to the next. The total weight (termed "yield") of the grass on some subplots was fifty percent higher than on others; height varied to a comparable extent. To test for genetic differences, they planted the seeds from different plots side by side (a true common garden experiment, in a real common garden!). Sweet vernal grass from the different plots grown under identical conditions in a university research garden differed in a variety of traits, including the weight of the flowers, the size of the leaves, and the grass's susceptibility to mildew, demonstrating a genetic basis for differences among the subplots.

The existence of evolved genetic differences among plots did not, in itself, prove that these changes were adaptive—the changes could represent the sort of random genetic fluctuations that occur by chance in small populations. To test the adaptation hypothesis directly, Snaydon and Davies grew plants under a variety of different soil conditions. As they expected, plants grew best on soil with the same chemical composition as their natal plot. Taking this approach one step further, they took garden-reared plants and placed them back out onto the experimental plots (by this point, the scientific dividends of the work were so obvious that the Plots Committee was more liberal in the sort of work it allowed). Sure enough, plants grew much better on their home plot than on plots with different soil chemistry and vegetation characteristics. The conclusion was clear: over the course of a century, plants had adapted to the conditions they experienced on their own subplots.

Snaydon followed up his initial research with additional studies, two of particular note. First, he and Davies looked at the boundaries between two pairs of plots, one in which the two plots had been receiving different fertilization treatments for 112 years, the other for 60 years. At both boundaries, they compared the plants on either side, only inches away from plants growing on soil with different chemistry. Later, Snaydon and another student, Tom Davies (unrelated to Stuart), looked at five plots that only six years previously had been split in half, one side newly treated with lime, the other remaining unlimed. In all cases, the results were very much in line with their initial findings. Populations evolved differences very quickly and over very short distances.

Snaydon and Davies were mainly interested in whether and how fast populations adapted. As a result, most of the data they collected and reported were not particularly directed toward the question of predictability of evolution, and extracting relevant information from the papers today, three to four decades after the fact, is not possible.

Nonetheless, in at least one way, Snaydon and Davies demonstrated that plant adaptation was not only rapid, but also highly repeatable. Because of the differences in soil composition, the height of all the plants varied substantially from one plot to the next. In turn, the sweet vernal grass adapted to this variation: on plots on which the other plants were very tall, sweet vernal grass itself grew higher and more upright—better for accessing the Sun's rays—and evolved greater shade tolerance than sweet vernal grass growing on plots with low vegetation.

It's always dangerous to state that a particular scientific paper was the first to propose a new idea or take a new approach. Someone will quickly point out that you overlooked some obscure reference predating the work you were touting. But I'll risk saying that the studies by Snaydon and Davies were the first to show that experiments could be used to study long-term evolution in the field.

The Snaydon and Davies papers on the Park Grass Experiment appeared from 1970 to 1982, exactly the time that the need for experimental approaches was being recognized in ecology and the rapidity of evolution was being appreciated in evolutionary biology. As a result, you might have thought that this work would have played a prominent role in uniting the two perspectives, a clarion call for the role of evolution field experiments.

It didn't. The papers certainly haven't been forgotten, but until recently they weren't widely known outside plant evolution circles—indeed, I was unaware of them until I started researching this book. When the papers are cited, it is usually in the context of plants diverging over short distances in response to different natural selection pressures, the phenomenon that Snaydon's advisor, Tony Bradshaw, had first established. Occasionally the rapid-evolution angle of the work is emphasized, but until recently, only rarely was it held up as an example of how evolution can be experimentally studied in a natural setting.

That's changed in the past decade. In 2007, an influential review of ecological experiments used to study evolution highlighted the Park Grass Experiment. Popular articles are now listing it alongside Reznick's guppy work. And currently molecular biologists are studying PGE's sweet vernal grass populations, looking to see whether the same genetic changes have occurred repeatedly as the grass adapts to new soil conditions. It took four decades, but the still ongoing work at Rothamsted has taken its place in the developing pantheon of field evolution experiments.

IT MAY HAVE TAKEN A LONG TIME for scientists to grasp the broader significance of Snaydon's work, but the same can't be said for Endler's and Reznick's studies. Their work clearly demonstrated that evolution could be studied experimentally in nature. Often in scientific research, once a novel method has been developed, a gold rush ensues as a herd of researchers embrace the approach, adapting it to address a variety of outstanding questions in the field. The guppy experiments were innovative and received a lot of attention, and were quickly followed by . . . very little. Our work on anoles—which started in a Snaydonian way by building on an ecological experiment, but then led to the initiation of brand-new experiments—was one of very few field evolution experiments conducted before the turn of the century. It wasn't until more than two decades after Endler's original work and three decades after Snaydon's that the experimental evolution stampede finally materialized.

Some of these studies followed Snaydon's lead by putting other long-running ecological field experiments under the evolution microscope. The most notable of these studies was conducted on another old estate, Silwood Park, just forty miles southwest of Rothamsted, where ecologist Mick Crawley had been excluding rabbits from small pasture plots for more than twenty years. Removing the rabbits had a huge ef-

fect on the vegetation: all one had to do was look on either side of the fence separating an exclosure from a rabbit-grazed control plot. Outside, the vegetation was very short, looking like a well-manicured field. Flowers were few, as were the seeds they produce. Reproduction was by vegetative spreading, a plant sending out a runner that sprouted up another plant. Inside the fence was a different world, with the appearance of an abandoned, unkempt lot. Plant life was wild and woolly. Flowers were abundant; seeds begat the next generation. As the years went by, the plots with rabbits maintained their front lawn trim, while the exclosures became increasingly disheveled. After five years, grasses that grow in tussocky clumps began to dominate; shrubs eventually took over in some of the rabbit-free plots. Given enough time, many of these plots turned into mini-forests.

But, just as with Rothamsted pre-Snaydon, no one had bothered to ask whether evolution was occurring, whether plants within the exclosures were adapting to their very different environmental conditions. Then along came Marc Johnson, a Canadian evolutionary ecologist now on the faculty of the University of Toronto. Every few years, Silwood Park researchers had started new experiments while continuing to maintain the old ones. The result was a series of plots in which rabbits had been excluded for different lengths of time. Johnson latched onto these plots and made a double-barreled prediction: not only would plants adapt to the absence of grazing, but also the degree of adaptation would increase with the number of years rabbits had been excluded from a plot.

One of Johnson's graduate students, Nash Turley, took the lead in the first part of the project. Turley looked at common sorrel, a slender herb sporting showy red fruits that is often grown for the tangy taste it imparts to a salad. Turley measured how rapidly the plants grew in the greenhouse and found a very strong trend: the longer a plot had been free of rabbits, the slower the plants grew. Over the course of a quarter century, growth rate evolved to be thirty percent slower in the absence of rabbits.

The success of that study led Johnson and Turley, with the assistance of stellar undergraduate Teresa Didiano, to examine other plant species. Three of the four species—all grasses—showed evidence of adaptation. For example, the number of leaves on red fescue decreased with the age of the exclosure. However, although three of the species had adapted to the lack of grazing, they did so in different ways, involving different traits. Moreover, the one forb, grass-leaf starwort, showed no consistent trend in any trait related to the age of the exclosures.

How you view the results of these studies depends on whether you're a glass-half-empty or a glass-half-full kind of person. In the Silwood Park studies, populations of the same species mostly evolve in a predictable way in response to being released from grazing, predictable in the sense that the longer their freedom, the greater the extent of their adaptation to a rabbit-free lifestyle. But when compared across species, the way they adapted was not predictable—different species evolved in different ways in response to the same environmental conditions.

AT ABOUT THE SAME TIME that Johnson and others were retrofitting ecological experiments, evolutionary biologists finally began en masse to set up field experiments explicitly designed to study evolution. These studies are diverse and fascinating. For example, Marc Johnson and colleagues at Cornell grew evening primrose on plots on which plant-eating insects had been removed with insecticide. The eight populations of herbivore-free primroses evolved in the same way over a three-year period, flowering earlier and putting fewer defensive chemicals into their seeds compared to primroses on the buggy plots.

Other studies have looked at how worms adapt in artificial enclosures with warmer soil (simulating climate change) or whether insects will quickly evolve camouflaged patterns if placed in small plots with different types of vegetation. Yet more studies are currently in progress.

However, it is the next stage—field evolution experiments on steroids—that is particularly exciting. No more popping lizards onto tiny unoccupied islands or using agricultural fields. Today's experimental evolutionists think big.

chapter eight

Evolution in Swimming Pools and Sandboxes

F ly over the southern end of the University of British Columbia campus in Vancouver and you'll see twenty aquamarine
rectangles. Arranged in four adjacent rows, their blue color
gives them away as full of water, deeper at one end and becoming more
shallow—evident by the paler shade of blue—at the other. Who plies
the twenty-fold cerulean waters of this natatory complex? Even Google
can't answer that question. Fortunately for us, however, Dolph Schluter
can. The lanky, always grinning but slightly shy Canadian may seem
more like a kindly organic farmer than a brilliant scientist, but Schluter
is the leading evolutionary biologist of his generation. These are his
pools, and their inhabitants are his charges.

   Schluter's upbringing gave no clue that he would become an aquatic
real estate mogul, designing an experimental evolution complex unri-
valed anywhere in the world. Always interested in nature, he supported
himself through college by assisting with field studies on Canadian
snapping turtles. As he prepared to graduate from college, he had a
job lined up performing mammal surveys in Alberta. Then, at the last
minute, he heard a brilliant research talk on hummingbird ecology

and realized that he wanted to be a scientist. Off he went to graduate school.

And not just to any Ph.D. program. Schluter began studies at the University of Michigan, where his doctoral advisor was none other than Peter Grant, the finch guru. Quickly, Schluter found himself in the Galápagos, putting his ideas into practice as he studied how Darwin's finch species adapted to use different resources. His detailed studies are now classics, appearing in textbooks and changing the way biologists study adaptive radiation.

But by the time he became a postdoc in Vancouver, Schluter was looking for a new study organism. Darwin's finches were great, but the Galápagos are a long way from Canada. More importantly, Schluter wanted to conduct experiments: to not only devise hypotheses from patterns in nature, but to also close the loop by testing them experimentally. Such experiments would be logistically difficult on any bird species and were not even possible on Darwin's finches due to the strict regulations of Galápagos National Park.

Fortunately, the answer was at hand. The little three-spined stickleback fish was the perfect solution: it exhibits interesting evolutionary patterns, is easy to study and manipulate in both the field and the lab, and is common in the lakes of British Columbia. At the time, stickleback were little known in evolutionary biology circles, but now—thanks in large part to Schluter's work—the fish has become a model organism for evolutionary studies.

Stickles, as apparently no one but me calls them, occur around the world in northern areas, but in several British Columbia lakes they do something that doesn't happen anywhere else. In most places, you can find only one three-spined stickleback species. But in five BC lakes, there are two species of stickles, a streamlined, fast-swimming species that lives in the open water offshore, and a second, chubbier, slower species that hangs out at the bottom near the shore. The species differ phe-

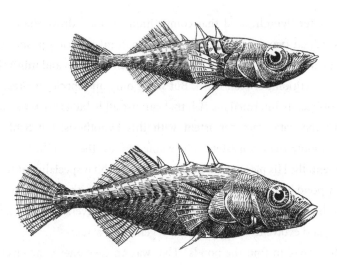

Open-water (top) and bottom-dwelling (bottom) three-spined sticklebacks

notypically. The open-water species has armor plates on its sides and long slender jaws that can quickly be protruded to nab open-water prey; conversely, the bottom-dweller doesn't have any armor and has powerful jaws to suck up prey from the sediment and attached vegetation.

Through DNA comparisons, Schluter's colleagues at UBC have shown that these two types—the open-water and bottom-dwellers—have evolved independently in each of the five lakes, the same pattern of repeated adaptive radiation shown by *Anolis* lizards in the Caribbean. In all other lakes, a single three-spined stickleback species uses both habitats and is more or less intermediate in body shape. Conversely, the open-water and bottom-dweller species are never found by themselves; they only occur together, in the five two-species lakes.

Schluter's growth and foraging studies, conducted both in the lab and in the field, showed that fish from one-species lakes are generalists, capable of living anywhere, but not excelling in any habitat. By contrast, in two-species lakes, the open-water and bottom-dweller species have each become specialized to use a specific habitat.

Schluter hypothesized that competition for food drove these patterns. When two species occur together, natural selection pushes the two species to diverge, specializing for different habitats and minimizing competition between them. But when only one species is present, phenotypically intermediate fish that can use all habitats are favored.

All the data were consistent with this hypothesis, but Schluter wanted more than consistency: he wanted to test the idea directly, experimentally. His plan was to place one of the habitat specialists into an empty pond by itself. According to his hypothesis, in the absence of the other habitat specialist, the population should reverse evolutionary course and evolve to the intermediate, generalist state.

But where to find the ponds? That was an easy one: Vancouver is full of man-made ponds lacking sticklebacks. Why not drop some fish into some of them? And so a trial run for the project began. Schluter received permission to put stickles into two ponds on golf courses and another in a municipal park. In went the open-water specialist species. Initially, things went well, but then, a year later, the golf course drained one of the lakes. The other two populations are still going strong to this day, but Schluter hasn't done much with them.

The reason is that shortly after he set up this experiment, UBC offered him a faculty position. Schluter reconsidered the golf course option. Wouldn't it be great if he could build a series of ponds, more or less identical in all respects, on the UBC campus? They'd be easy to access and safe from interference, and without the risk of being bonked in the head by an errant stroke.

The university gave the approval, a contractor was hired, and in went the ponds, all thirteen of them, each seventy-five feet on a side, gradually sloping to a depth of ten feet at the center. The ponds were initially inoculated with plants and insects from a nearby two-stickleback-species lake and then were left to their own devices. Trees—a veritable forest after a few years—grew up around the edges, birds arrived; eventually, the ponds looked completely natural. At times you could forget you were just across the street from the UBC campus.

Over the course of seventeen years, Schluter and his lab used the ponds to measure how natural selection operated on the sticklebacks, investigating which traits led to better survival and why hybrids between the two species were at a selective disadvantage. The work was wildly successful, enshrining sticklebacks as a textbook case of evolutionary divergence driven by competition for resources. Still, most of the work was within a single generation, measuring survival and reproduction, but not its multigenerational, evolutionary consequences. Finally, the time came to try an evolution experiment.*

To initiate the study, star Schluter lab graduate student Rowan Barrett—more on him shortly—collected marine sticklebacks from a nearby lagoon. Lake sticklebacks are derived from marine sticklebacks that became trapped inland when the British Columbia landmass rose after the glaciers melted at the end of the last Ice Age. Marine sticklebacks, which have more armor plates and are adapted to less extreme temperatures, thus are similar to the ancestral condition of the lake populations.

Barrett placed marine stickles in three of the experimental ponds to test whether they would adapt to freshwater living just as descendants of marine fish had done in real lakes. The results for the armor plates were complicated and ambiguous. But the experiment investigated a second trait, one relevant to present-day concerns: how rapidly can sticklebacks adapt to changing climate conditions? Water temperature in freshwater lakes is more variable than in the ocean—hotter in the summer, colder in the winter. Would the marine sticklebacks adapt to these broader extremes?

---

* Actually, this was the second attempt. Shortly after constructing the ponds, Schluter had tried an evolution experiment, but it had failed, leading him to focus on within-generation studies of natural selection.

To find out, Barrett recorded a standard measurement in thermal biology, the high and low temperatures at which fish lose their ability to move in a coordinated way. When examining marine and lake fish, Barrett found that they did not differ in the maximum temperature they could tolerate—for some reason, sticklebacks can handle temperatures much warmer than it ever gets in their habitats. But low-temperature tolerance was different. Lake fish can function at temperatures five degrees Fahrenheit lower than ocean fish, a difference that almost precisely matches the differences in cold experienced by fish in the two environments. Consequently, Barrett focused on cold adaptation—would the experimental pond fish evolve to become more cold-hardy?

The answer was yes, and very quickly. The cold winters took their toll and those that couldn't handle it died; after just two years, fish in all three ponds had already evolved the ability to tolerate temperatures four and a half degrees lower than their marine ancestors, almost matching the cold hardiness of British Columbia lake sticklebacks.

This remarkably fast and parallel adaptation was not expected, and Barrett, Schluter, and company were excited to see what would happen next. Unfortunately, they didn't foresee the next development, either. The winter of 2008–2009 was the coldest in four decades, and the challenge it posed was too great for any of the fish to handle. All perished, turning the study into a shorter long-term experiment than planned. Nonetheless, the study clearly demonstrated the ability to conduct evolution experiments in the experimental pond complex.

All good things must come to an end, and so it was with Schluter's experimental ponds. Because of the porosity of the soil, the ponds had been lined with plastic sheets—otherwise, the water would simply drain away. Schluter had been warned from the outset that the plastic had a twenty-year lifespan and the expiration date was rapidly approaching. It had been a good run, but it wasn't clear what came next.

Sometimes good things happen to good people. Out of the blue, Schluter received a phone call from a campus official. The university wanted its land back to build a new housing development that it could sell for exorbitant amounts in the Vancouver real estate market. Would Schluter be willing to move his operation elsewhere if the university helped him build new ponds?

To ask the question was to answer it, and Schluter got his new state-of-the-art stickleback pond complex, a stone's throw away from where the original ponds had been (and where deluxe condos, a craft beer pub, a music school, and restaurants now reside). The ponds are about the same size as the old ones, but differently configured: rectangular, rather than square, and sloping to a depth of twenty feet at one end to more closely resemble the natural lakes.

It took several years to build, but the Speciation Accelerator (as Schluter likes to call it jokingly in reference to the cyclotron at the subatomic physics laboratory across the street that accelerates charged particles to very high speeds) is now up and running. The first multi-generational study has already been completed, examining the role of predators on the evolution of defensive traits. Five ponds were stocked with stickles and predatory cutthroat trout, while in five others the sticklebacks were left by themselves. The experiment was allowed to run for five generations.

Around the same time that the new pond complex was being built, Schluter embarked on a new research direction: the field biologist became a geneticist. Working with leading genome experts at Stanford and elsewhere, Schluter was involved in getting the genome of the three-spined stickleback sequenced, leading to the identification of the genes underlying key traits, such as the possession of armor and spines.

As a result of this newfound genetic knowledge, the pond experiment had a two-pronged approach, examining evolution at both the phenotypic and genetic levels. The prediction was that the presence of predators would drive the evolution of longer spines, which make the fish harder to swallow, as well as the genes responsible for spine length.

And the results? Graduate student Diana Rennison has only analyzed the data through the first three generations, but the results are very promising. Looking at survival within a generation, fish with longer dorsal spines survive better in the ponds containing predators. This selection has led to an evolutionary response: the predator-pond populations now have longer dorsal spines. Results at the genetic level have been parallel, with gene variations causing longer dorsal spines increasing in frequency in those populations. Curiously, though, selection for pelvic spine length has been more variable, with longer spines being favored in only some of these ponds—the cause of this evolutionary indeterminacy is unknown.

It's still early days in stickleback experimental evolution, but already the results are very similar to what we've seen with guppies. Populations adapt to new circumstances mostly in the same way, yet with some degree of unpredictability in some traits. The concordance is particularly compelling given the differences in the settings in which the fish have been studied, one in natural streams in the mountains of Trinidad, the other in nearly identical artificial ponds in Vancouver.

JUST AS SCHLUTER'S PONDS WERE FILLING, another colossal experimental evolution study was beginning to take form in the American interior. Like a Christo installation in steel, nearly half a mile of sheet metal now adorns the Nebraska landscape, shimmering in the summer heat, glowing orange in the setting sun's reflection. The American West is crisscrossed with fences of all sorts, but this one is unique, a solid metal wall in the shape of a square, subdivided into four square compartments. And not just one, but two such structures, almost exactly thirty miles apart on different-colored soil.

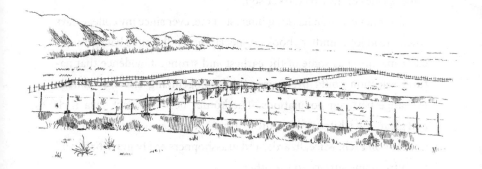

The rolling hills and prairies of Nebraska are known for their rich, fertile land—brown, earthy, packed full of plant goodness. There's a reason the name of the state university's football team is the Cornhuskers. But not all of the state's land is so productive. About a quarter of the state is in the Sandhills region, an area in which the soil is sandy and light-colored, composed of quartz specks blown east from the Rocky Mountains about eight thousand years ago. Crops grow poorly there and most of the area has never been cultivated.

That's not to say that nothing flourishes on the Sandhills. Quite the contrary, the area is biologically rich, so distinctive that it's recognized as its own ecoregion by the World Wildlife Fund. The soil affects the region's biota not only by dint of its low fertility, but also as a result of its light coloration. All around the world, small animals have evolved to match their background, all the better to avoid being detected by their predators. On old lava flows, lizards, mice, grasshoppers, and other animals have evolved to be much darker than they are elsewhere. Conversely, on light-colored soil, animals evolve a pale complexion to blend in with the sandy substrate. The Nebraska Sandhills are no different—

populations of many species there are lighter than members of the same species on nearby darker soil.

This phenomenon has long interested me, ever since my college days when I read John Endler's book on speciation. Endler cited background-color matching as some of the earliest and strongest evidence that natural selection could overpower the homogenizing effect of genetic exchange between populations. The border between black lava rock and gleaming white sand is quite distinct—in places you can stand with one foot on each. Mice, lizards, and grasshoppers easily move back and forth from one surface to the other.

Yet, despite their proximity, populations on the different surfaces are often very different in coloration, appropriately matched to where they live. Members of the two populations can meet near the border, but any offspring from such encounters are scrutinized by natural selection and genes for the mismatching color quickly weeded out. The fact that some of these environments—the Nebraska Sandhills, some lava flows—only appeared recently indicates that color adaptation has occurred rapidly, providing further evidence for the power of natural selection even in the presence of gene flow. These are the animal analogs to plants living on or off old mine sites or on the different Rothamsted plots.

Research on deer mice—so named for their running and jumping agility—has been particularly influential. Naturalists in the middle of the last century noted many cases in which neighboring populations occurred on different-colored substrates and had correspondingly different fur coloration. The presumed explanation was camouflage—rodents are eaten by many visually oriented predators, so natural selection would drive populations to become similar in coloration to the background on which they occurred.

A University of Michigan biologist, Lee Dice, even tested the idea in the lab. Dice took a normal-sized room, covered it with soil, and released different-colored mice. And an owl. His goal was to see whether the bird more frequently captured mice that didn't match the soil. Half the trials were conducted on light soil, the rest on dark soil. The answer was clear-cut: the owls caught twice as many mismatched mice as those that blended in. Bird predation could indeed be a strong agent of natural selection, propelling the evolution of camouflage.

Nonetheless, this was a laboratory study conducted in highly contrived conditions. Even thirty years after Endler's book was published, the operation of natural selection hadn't been demonstrated directly on mouse coloration. In fact, the strongest evidence had come not from field studies, but from genetic investigations that had discovered the genes responsible for differences in coloration in mice. By comparing DNA differences in adjacent populations on different-colored substrates, researchers discovered that genetic differences had evolved very recently, probably the result of divergent natural selection pressures. But this, still, was inference drawn from genetic differences, not direct demonstration of natural selection causing evolution.

And that's what brought a bearded Canadian ski bum with the ropy build of a cyclist to the Nebraska Sandhills. Despite his ardor for backcountry skiing, cycling, and rock-climbing, Rowan Barrett spends most of his time studying evolution. The son of an eminent evolutionary biologist at the University of Toronto, the younger Barrett has blazed his own trail; only in his mid-thirties, he has become one of the leading figures in experimental evolution. As a master's student at McGill University in Montreal, Barrett conducted experimental studies on bacteria in the lab, studying how they adapt when confronted with multiple new resources. He followed that with the doctoral work with Schluter at UBC that I've just described. These studies were extraordinarily successful, leading to a string of publications in high-profile journals and landing Barrett every possible prize awarded to outstanding young evolutionary biologists (in essence, he was named to the evolutionary biology All-Star team and won both European and North American Rookie of the Year awards).

But this was all just a prelude for what promises to be his biggest hit. Toward the end of his doctoral work at UBC, Barrett learned of the genetic studies in the laboratory of my colleague at Harvard, Hopi Hoekstra, that demonstrated that Sandhills deer mice had evolved a new mutation that produced light coloration. Comparison of the DNA of these mice to those of nearby dark-colored mice suggested that the mutation had arisen recently and had swept through the population, probably under the power of natural selection for background matching.

Barrett, however, felt that the story was incomplete. If natural selection was responsible for the evolution of light color, then it should be possible to demonstrate it directly. And his inner experimentalist told him just how to do it: take the approach Dice had pioneered seventy years previously. Put dark and light mice on dark and light substrates and see which ones survive. Only don't do it in an enclosed room; do it out in nature. And just like the stickleback experiments, study both the phenotype and the genes responsible for producing it.

It sounds easy enough in principle, but how do you put a plan like

Sandhills deer mouse

this into practice? The advantage of ponds, streams, and islands is that they're self-contained units with hard boundaries, experimental replicates just waiting to be used. Nothing comparable exists in the Sandhills. If Barrett wanted to conduct deer mouse experiments there, he was going to have to build cages to keep his subjects where he wanted them. Enormous cages, big enough to hold entire populations of deer mice.

Researchers had tried similar experiments in the past, on a smaller scale. They had always failed. Even though deer mice live on the ground and in burrows, they are very nimble. Put up a wall and they will climb it. Ninja mice, some have called them. Previous studies had been foiled by deer mouse jailbreaks. Barrett's first task was to find a way to keep the mice in their cages.

After doing some sleuthing, Barrett discovered that the problem had already been solved. Ecologists and mammalogists hadn't figured out how to contain deer mice, but disease researchers had. Deer mice are carriers of hantavirus, prompting scientists to devise escape-proof outdoor cages in New Mexico where mice could be quarantined long enough to assure they were hantavirus-free so that they could be sent to research labs for studies. The trick was 26-gauge galvanized steel, smooth as the proverbial baby's bottom, offering no irregularities that an adventurous deer mouse could latch onto with its claws and so maneuver its way up and out. In trials back at the lab, Barrett confirmed that the mice were stymied by the thin metal. He had his plan.

Still, there were two big challenges: getting permission to build the cages on appropriate land and then actually constructing them. The dimension of these issues becomes apparent when you consider the size of the cages Barrett planned to build. He figured that each deer mouse population should contain at least one hundred mice; based on natural density levels, that translated into a bit more than half an acre. For good experimental design, he wanted to have four cages on light soil and four on dark, requiring two and a half acres of land and fifteen thousand pounds of sheet metal on each site.

But first things first. Barrett, having joined the Hoekstra lab as a postdoctoral fellow, headed out to Nebraska to scout for study sites with Catherine Linnen, the postdoc who had performed the genetic studies in Hoekstra's lab. Finding a place with suitable light-colored soil was not too difficult and arrangements were quickly made to build cages on the grounds of the Merritt Reservoir wildlife refuge area.

Finding land on dark soil was more difficult. Dark soil, after all, is fertile soil. Convincing someone to put aside two and a half acres of good cropland so that a cage could be built on it to hold mice was not going to be an easy sell.

Barrett went house to house, talking to landowners. Remember that this is the middle of the American heartland, a politically and religiously conservative region where people make their living by tending their land, producing the food that the rest of the country lives on. And now look at who's knocking on the door. A couple of kids, not only from the liberal Northeast, but from some elite, effete Ivy League school. One of them wasn't even American, and he went around wearing a biker's short-billed cap instead of a cowboy hat.

Barrett quickly learned not to mention the e-word—evolution—but rather to talk about species fitting into their environment. Being farmers and ranchers, the locals understood heredity very well and knew all about predators; moreover, the idea of camouflage was second nature to folks who had been hunting since they were kids. And Barrett's outgoing, an easy conversationalist. People were very friendly, even interested.* They were happy to let Barrett, Linnen, and the rest of the team

---

* This reminds me of a story John Endler once told me, about being on an airplane and reading a book on speciation. The man in the next seat asked what he was reading, and they launched into a conversation in which Endler explained all about natural selection, evolution, and speciation, but without using any of those terms. The man became quite inter-

collect mice on their property. But as for turning over several acres of land, that was asking too much.

This continued well into the field season and time was running short; Barrett was beginning to despair that the project would have to be put on hold. One night in the small town of Valentine, Nebraska (population: 2,737), Barrett went to the hotel bar for a beer, and the owner introduced him to Wild Bill, a local celebrity whose long blond hair makes him look more like a surfer dude than a Nebraska farmer. More making conversation than a pitch, Barrett told Bill about the project and the quest for a study site. To Barrett's surprise, Bill said that he grew alfalfa on land outside of town and perhaps the cages could be built there. The next day they went for a look and the land was perfect. Wild Bill was unconcerned about the construction that would be needed, and as for rent, a case of Miller Lite and a barbecue each time Barrett's team came to town would do the trick.

Of course, finding the sites was only the first step. Next the enclosures had to be built. Barrett is from an academic family; he had no construction experience. And you can't look up "mouse-enclosure construction" in the Nebraska Yellow Pages. Barrett had to figure it out himself.

He studied other enclosures researchers had built and settled on a design. The walls had to be three feet high to keep the mice from jumping out and topped with another three feet of chicken wire to dis-

---

ested, followed along attentively, and asked some very good questions. Finally, he asked Endler for reading recommendations to learn more, to which Endler started to respond that the best place to begin was with Darwin. But the moment that name left Endler's lips, the man became red in the face, turned away, and didn't utter another word for the rest of the flight.

In a similar vein, the National Science Foundation requires researchers to write a short summary of funded grants to release to the public. At one time not too long ago, evolutionary biologists were advised to describe the work without using the word "evolution." Apparently, many opponents have no problem with the basic ideas of natural selection and descent with modification as long as they're not labeled with the e-word.

courage coyotes from jumping in. Below the soil, the wall needed to extend down two feet, deep enough that the mice couldn't burrow below and make their escape.

Few evolutionary biologists have their research equipment delivered by flatbed semi-trailers. A truck from Kimball, Nebraska, 250 miles away, delivered the steel walls in five-foot-by-ten-foot sheets, one-fiftieth of an inch thick. A rented Ditch Witch dug the trench into which the sheets were placed and a local backhoe operator moved the sheets from the road to the fields. One hundred ninety-two posts were sealed into the ground with concrete, supporting the wall at the junction between adjacent metal sheets. The whole operation took two weeks and involved three local construction workers to operate the machinery, four groundskeepers from the local golf course to help dig trenches, and seven lab members to do the rest (you know the old joke—how many scientists does it take to build a mouse enclosure?).

Perhaps it's a credit to Barrett's careful management, perhaps to the hard work of the Harvard lab members who were unfazed by being out of their element, perhaps the local equipment operators deserve double credit for getting the job done in the presence of such tenderfoots. Regardless of who gets the kudos—probably some combination of all three—the work went remarkably smoothly and quickly. Sure, there was the time that the backhoe almost rolled over after being front-loaded with too many steel plates, and the time high winds sent razor-sharp, fifty-pound steel rectangles flying through the fields, but no harm was done, and at the end of two weeks, the enclosures were ready for mouse move-in day.

The mice, however, were not. They were still living in their burrows, unaware of the fate that was soon to befall them. Barrett's plan was to seed each enclosure with an equal number of light-colored and dark-colored mice. But to do that, he and his team first had to capture the mice.

The time-honored way to catch live rodents is to go out into the field late in the afternoon and place on the ground a large number of foot-long metal boxes with one open end. Inside the box is the bait— yummy seeds, peanut butter, or something else tasty—and a horizontal platform. When the mouse (or other creature, anything from a scorpion to a snake) steps (or slithers) on the platform, the trapdoor slams shut and the animal is trapped inside. Then come back early the next morning, pick up the traps, and carefully peek inside to see what you've caught.

Barrett and company had been trapping mice in Nebraska for quite some time at this point, collecting samples from throughout the region for their genetic studies. Based on their success rate, Barrett anticipated no problem in quickly getting the mice he needed.

The mice, however, had other ideas and started shunning the traps in unprecedented numbers. One night, the team put out seven hundred traps and came back the next day to find just two mice (a more typical yield would have been thirty-five). What should have taken a week or two took three months. But finally the enclosures were stocked and the experiment under way.

Barrett had one last decision to make. The experiment was about the effect of visually oriented predators on mouse coloration, but not all predators find their prey by sight. Some use smell or even heat. These predators should capture mice randomly with respect to their coloration; such predation would add unpredictable noise to the results, possibly, just by chance, obscuring the effects of visual predators. Perhaps they should be excluded to avoid this possible pitfall. On the other hand, this was meant to be an experiment in a natural system, and these predators are a part of that system. Barrett waffled on what to do.

He was particularly concerned about snakes. The prairie rattlesnake is common in the Sandhills. Attaining a length of nearly four feet, the adults prey on a wide variety of mammals up to the size of a prairie dog; deer mice are just the right size for growing youngsters. Bull snakes can grow to twice the size of prairie rattlers and have a particular fondness for rodents. Barrett decided to remove them— carefully—from the enclosures. Every snake encountered was gently lifted with a snake-catching stick with pincers on the end and released outside the enclosure.

But they kept finding more snakes. How many could there be living in a two-and-a-half-acre field? When would they finally have caught them all? And then they realized: the walls weren't keeping the snakes out. Given their lack of legs, snakes are surprisingly good climbers, and a three-foot metal wall is not much of a challenge for a large prairie rattler or bull snake. As fast as Barrett could catch and remove them, more (or perhaps the same ones) were getting back in. The snake removal ceased. This would be a completely natural experiment after all.

Once the mice were in the enclosures, all Barrett could do was sit back and wait for evolution to happen. Every three months, he'd lead a team back to Nebraska to census the enclosures. Live traps were laid out in the enclosures to see who was still there. When they were introduced into the enclosure, each mouse had been injected with a small tag, the same kind people put in their dogs and cats, with an individual barcode. Run the scanner over the mouse and if it was one of the originals, its ID number would pop up on the screen. In ten days, almost every mouse in every enclosure could be captured, scanned, and released.

The return visits were also a time to reconnect with the many friends Barrett and his team had made in Valentine, which had turned out to be as welcoming as its name suggests. One family would have the crew over for dinner almost every night. An elderly couple let some team members stay in their house for next to nothing and provided their garage for equipment storage. On every visit, Barrett would throw a big party for one and all.

And the mice did their part as well. At the start of the experiment, mortality was very high, not surprising when you let animals go in a new place. While they're still looking around, getting familiar, they're easy pickings for predators. In addition, as we all know, there's a lot of stress involved in moving into a new home—especially when the move is involuntary—which no doubt contributed to the high mortality rates.

But what was exciting was not the level of mortality, but who made it and who didn't. In each of the enclosures on the sandhills, the light-colored mice survived better than their darker cousins, by a two-to-one margin on average. Conversely, in the dark-soil enclosures, the tables were turned and it was the dark mice that were the champions, surviving at a rate a third higher than their lighter compatriots. Examination of the mice's genotypes provided comparable results. On the light sand, individuals with mutations producing light coloration survived better, whereas on the dark soil, the situation was reversed. Natural selection was operating, and in different directions in the different enclosures, just as predicted.

Fifteen months into the project, all the original mice were gone. But the populations were doing just fine, composed of the offspring of those founding individuals. The selection experiment had now become a long-term evolution experiment.

As I write this, the experiment has now been ongoing for five years, enough time for about ten deer mouse generations. Barrett is just now pulling together all the results, finishing up the genetic analyses. He doesn't know yet what the results will show, but if the strong selection at the outset is any indication, it likely will be that the populations have evolved in opposite directions. Still, nature is full of surprises, so Barrett is keeping an open mind. The paper will probably be finished about the same time as this book—given the experiment's grandeur, look for the results to be reported in the *New York Times*.

EXPERIMENTAL STUDIES OF EVOLUTION in the field are only going to get larger, bolder, more exciting. Already, one study is pumping a constant stream of carbon dioxide onto a large field, raising levels to those projected for the global atmosphere fifty years into the future. Will plants evolve and if so, how?

Twenty years from now, perhaps even sooner, we will be awash in data from evolution experiments. Perhaps as more data come in, the story will change, but from what we know now, the general results are clear. When an experiment is set up in which multiple populations experience the same environmental treatment, the populations tend to evolve in very similar ways. This outcome is as true for plants safeguarded from marauding rabbits as it is for lizards forced to use narrow perches. Simon Conway Morris should be pleased. Evolution is repeatable.

This result should not be completely surprising. In the discussion of convergent evolution in Part One, I noted that closely related populations or species tend to evolve in the same way because they are similar genetically; selection has the same genetic material to work with and thus tends to fashion the same solution. In contrast, distant relatives, starting from different initial genetic constitutions and phenotypes, are more likely to evolve different adaptive responses to the same environmental challenge. Field evolution experiments always start with very similar populations, usually individuals drawn from the same source population. As a result, the experiments are predisposed to generate parallel evolutionary responses.

This is not to say the evolutionary change is identical from one experimental population to the next. Quite the contrary, there is always some degree of variation. In Nash Turley's rabbit-exclusion study, for example, among exclosures created at the same time, plant growth rate often varied from one population to the next. Among four six-year-old plots, for example, growth rate was fifty percent higher in the fastest-growing population than in the slowest. Similar variation occurred among the nine-, thirteen-, and twenty-five-year-old populations. Despite being released from rabbit grazing for the same amount of time, the populations varied quantitatively in their evolutionary response. Similarly, in the Cornell experiment in which evening primroses were sheltered from insects, plants began flowering much earlier in the season, but the extent of the evolutionary shift varied greatly among plots, with some producing five times more flowers early in the season than others.

Such variation may indicate a degree of indeterminacy in evolutionary response, even in closely related populations experiencing the same selective environment. Like most scientific research, experimental evolution studies focus on general trends, analyzed in a statistical framework. They tend to overlook exceptions; the occasional aberrant population—one adapting in a different way—may occur, but not be noticed. Papers often don't even report the raw data, so such outliers—those few that have embarked on a different evolutionary course—may not even be apparent to readers. The extent to which a minority of populations takes different paths is thus often unclear.

Furthermore, studies often measure many different traits, but only highlight those that evolve in similar ways. Those that evolve in many different ways in different populations won't exhibit statistically significant trends and thus are ignored, even though they may be evidence of divergent patterns of adaptation.

Of course, the existence of variation in population responses exposed to the same conditions is not necessarily evidence for evolutionary non-determinism. Another possibility, one that many would find more parsimonious, is that the environments the populations experienced may not have been exactly the same. Couldn't variation in plant traits reflect adaptations to subtle differences among the plots in soil composition, number of snails, or shading from trees? Might not differences in lizard limb length reflect slight differences in the species of bushes on different experimental islands?

We can't know. The great advantage of field experiments is that they are conducted out in nature, exposed to all the disparate selective factors that impinge on the real world. They are not an abstraction of nature or a simplification of it—they are truly representative of what populations face in the great outdoors.

But field experiments have one big disadvantage—you can't control for everything. Nature is varied, even over short distances. And those differences can confound the interpretation of results. That's why laboratory scientists shudder at the thought of doing experiments in the field—the lack of control gives them the willies. If you really want to know how repeatable evolution is, how much the same selective environment will predictably yield the same evolutionary outcome, then conduct your experiment in the lab, where the environment can be precisely controlled. Such studies will trade relevance to the real world for experimental rigor, but to thoroughly test Gould's postulate, that may be an exchange worth making.

Part Three

EVOLUTION UNDER THE MICROSCOPE

chapter nine

Replaying the Tape

The iconic localities of evolution: the Galápagos Islands. Olduvai Gorge. Australia. Madagascar.

East Lansing, Michigan?

Surprisingly, some of the most important research on evolution in the last several decades has come from studies on evolutionary change occurring in the middle of the Great Lakes state.

Room 6140 in the Biomedical and Physical Sciences Building of Michigan State University seems like any ordinary biology laboratory. Two high blacktop tables with shelves running down the middle partition the lab, creating three aisles. The sides of both tables house workstations where researchers sit, surrounded by the accoutrements of lab science: bottles full of amber and clear chemicals on the shelves, stacks of petri dishes, odd-looking pieces of tabletop equipment. The walls are bedecked with the usual assortment of postcards, nerdy science cartoons, and photos of animals and scientific celebrities. An inexplicable piece of wood hangs from a shelf, suspended horizontally by two bent paper clips. Small children's toys and other knickknacks peek out of corners and from behind computer monitors. A glass-doored deli-style refrigerator full of chemicals and other large machines line one wall. The other wall is composed of large windows stretching to the ceiling, looking out over the campus.

Many labs have their idiosyncrasies, and so does this one. Taped across several windows are pieces of blue paper, arranged to spell out in very large numbers, one per window pane, "000,ϞϿ." At least, that's what it appears to spell when viewed from inside the lab; from the sidewalk, outside and six floors below, the string of numbers, read in the opposite direction, is more sensible. We'll come back to that a bit later.

The lab bustles with scientists, mostly young, dressed casually in T-shirts and jeans. But one young man stands out—he's wearing a colorful garment in multiple shades of blue that appears to be a cross between a tie-dyed lab coat and a wizard's robe. We'll come back to him in due course, too.

The focal point of the lab—indeed, the lab's raison d'être—is a blocky piece of equipment next to the door. In size and appearance, it is similar to the horizontal freezers in gas station mini-marts that hold ice-cream sandwiches, Klondike bars, and the like. It's certainly newer-looking and more high-tech than what you'd find at a service station, but it still wouldn't have been surprising to find a stack of ice-cream drumsticks inside.

Finally, the moment I'd been waiting for arrived—my host lifted the faux freezer's lid, letting me see what was inside. I leaned forward and warm air wafted over my face—definitely no fudgesicles in this metal chest. In their place were two rows of seven small glass flasks, each snugly fit into its own holder on a metal plate. This metal platform slowly moved back and forth, right and left, gently undulating the small amount of liquid in each flask.

I have to admit to surprise, and perhaps a bit of disappointment. Here I was at ground zero of one of the most important studies in evolutionary biology in the last quarter century, and it was so . . . humble.

Pedestrian. Unimpressive. Just small containers of clear liquid, slowly sloshing to and fro in a tropical ice-cream bin.

THE STORY OF HOW SUCH small containers made such a big splash begins not in a Michigan laboratory, but nearly forty years ago in the Appalachian Mountains of North Carolina. There, a young graduate student named Rich Lenski was using the time-honored pitfall trap method to census beetle populations. Pitfalls work just like they sound. A researcher digs a hole and waits for animals to fall into it. Animals aren't so stupid, you might think. But it turns out they are. They go wandering along and—plunk—suddenly they've dropped into it and can't get out. The size of the pit depends on what you're trying to catch: for beetles, paper-cup-sized will do, but lizards and snakes require large buckets.

Lenski had long been fascinated by the natural world. The North Carolina native majored in biology at Oberlin, a small college in Ohio famous primarily for music, but also a great place to learn science. There, he became enamored with the experimental approach to studying scientific questions. So prevalent in laboratory research, experiments then were much less common in studying the natural world.

At just this time, the field of ecology was in turmoil. Most studies were observational and comparative: collect detailed data at multiple sites, then look for associations among variables to explain the similarities and differences. Perhaps localities with more butterflies also have more dragonflies. That might suggest that butterfly abundance determines how many dragonflies can occur at a given spot, which would make sense if dragonflies eat butterflies. But the causality might go the other way: perhaps dragonfly abundance determines the number of butterflies—dragonflies might eat butterfly predators, as one possible explanation, so more dragonflies would translate into fewer butterfly predators and thus more butterflies. Or maybe there is no direct link, but rather, the abundance of both is determined by a third variable: perhaps wetter sites have a positive effect on both butterflies and dragonflies. In that case, even if butterflies and dragonflies don't affect each other, their abundance would be correlated because they both respond to moisture levels.

This is the age-old problem of correlation versus causation and the most direct way to get at it is to conduct an experiment. If the size of the butterfly population determines the number of dragonflies, then altering the number of the former should lead to a change in the number of the latter. The experimental design is obvious: go to some sites and augment or reduce the number of butterflies and see how dragonfly populations respond. Of course, any change you see may be a fluke, perhaps driven by weather or some other extrinsic variable. To rule that out, you'll need a control site, one at which you do everything the same except that you don't change the number of butterflies. If you get a response in the experimental plot and not the control plot, you've got evidence that butterfly numbers affect dragonfly abundance. Actually, though, one pair of sites is not enough—any difference could still be a random fluctuation. Better to use a bunch of sites for butterfly population manipulation and another bunch for controls to make sure any trends are consistent. Of course, you'll need to randomly assign the sites to one treatment or the other to avoid any bias that might affect the results.

In the late 1970s, a cadre of ecologists, including Silwood Park's Mick Crawley, was pushing the experimental approach, arguing that decades of observational studies of ecology were misguided and that a more powerful method was needed. Conducting experiments in nature might be difficult, but doing so was necessary, they argued. One of the leaders in this movement was Nelson Hairston Sr. at the University of North Carolina, and it was to his lab that the young Lenski was drawn after graduating from college at age twenty.

Hairston was ecumenical in his interests, both in terms of questions to be addressed and organisms to be used, with the one constant that an experimental approach was paramount. Lenski's interest in how ecosystems function was certainly appropriate, and the beetles of Appalachia were a good study system.

Lenski's doctoral work focused on the abundance of two common beetle species in the forests of North Carolina. Lenski had two questions: do the species compete for resources and, revealing his strong interest in the environment and how humans are affecting it, how are beetle communities affected by forest clear-cutting?

Lenski approached these questions in two stages. First, he took the classical, comparative approach. He visited multiple sites and collected the beetles. With these data, he could see what factors were correlated with the abundance of each species. Then, to test hypotheses suggested by these data, he built large enclosures to conduct experiments in which these factors (beetle density, forest versus clear-cut) were altered.

The approach sounds simple, but it was an enormous amount of work. A single pitfall trap consisted of a hole dug four and a half inches deep, into which a plastic cup was then placed. That doesn't seem too bad. Except that for one study, 192 such holes were dug; for another, concurrent study, 64. Every day for two months, Lenski tromped up the mountain to the study sites and then went from one cup to another, checking the contents, removing, examining, and liberating anything inside.

Setting up and running the experiments was also laborious. Square enclosures—five feet on a side in one experiment, twenty feet in another—were constructed by sinking aluminum flashing into the ground. Beetles within them were monitored regularly with pitfall traps, the lucky ones hand-fed every time they were captured to see if their size and reproduction were limited by food availability. Some experiments lasted two weeks, others three months.

The two parts of the project meshed well, the experiments generally confirming the hypotheses suggested by the observational studies. The results showed that beetles in the forest did better than those in the clear-cut and that competition for food between the species may be an important factor regulating their populations.

As doctoral theses go, Lenski's was a big success. It led to six published papers, three in the very top journals in the field, indicating both that this was high-quality work and that Lenski was an up-and-comer. The future was very rosy.

Yet all was not well—Lenski was not thrilled with his research program. His interests had changed during his years as a graduate student. A series of thought-provoking courses on evolution, as well as innumerable discussions with like-minded grad students over games of Pac-Man and Friday night beers, had sparked his interest in studying how organisms adapt to their environment. And so he set out to find an organism more suitable for studying evolutionary change, particularly one amenable to the experimental approach he so clearly appreciated.

In graduate school, Lenski had read about a classic experiment on the genetics of microbes.* "I figured that as long as I was going to work with something new and unfamiliar (pretty much everything), microbes had the advantages of a model system that had proven its worth in other fields," he says. Thus the coleopterist became a microbiologist.

LENSKI WAS FOLLOWING A LONG LINE of scientists who had studied evolution in laboratory populations. Since the early years of the twentieth century, thousands—perhaps tens of thousands—of such experiments have been conducted. And the results have been surprisingly consistent. For any organism that can be brought into the lab and bred, selection on almost any trait will lead to a rapid evolutionary response in the predicted direction. These studies have included not just the traits you'd expect—body size, color, the number of bristles on the butts of fruit flies—but a wide variety of other traits, such as the resistance of rats to developing cavities, the tendency of fruit flies to fly toward light, and fruit fly tolerance of alcohol fumes (more on that in Chapter Eleven). For the most part, pick any trait that varies in a population, impose artificial selection, and you will get an evolutionary response.

Essentially the same approach has been used to produce the farm animals and agricultural plants that are familiar to us, but bear little resemblance to their wild ancestors. For example, teosinte, the progenitor of corn from the highlands of Mexico, has an ear four inches long with perhaps a dozen kernels and a tough outer covering—that's a long way from the hundreds of luscious, unprotected kernels in the cobs we

---

* The 1943 paper "Mutations of Bacteria from Virus Sensitivity to Virus Resistance" demonstrated that bacterial heredity was gene-based just as in plants and animals and that mutations occur randomly. Salvador Luria and Max Delbrück were awarded the Nobel Prize for this work.

feast upon every summer. Factory chickens can produce more than three hundred eggs per year, vastly more than their jungle fowl ancestors. In the same way, artificial selection has produced the great dane and the chihuahua from a primordial wolf.

Artificial selection has been a boon to science, to agriculture, and to human well-being. But as an analog to the natural evolutionary process, it is incomplete. Lenski, just finishing his doctoral work, realized two ways that laboratory studies could be made more similar to what goes on in nature.

First, when we think of evolution, we think of what goes on over thousands and millions of years. Yet laboratory selection studies generally only lasted a few tens of generations—enough to see a strong evolutionary response and to learn a lot, but still far short of nature's timescales. And the reason, of course, is obvious: human scientific careers aren't that long, much less the grant cycle over which a scientist needs to get results and write papers in order to get the next grant. Moreover, the organisms used in such studies—fruit flies, mice, and the like—have generation times of many weeks to months, limiting the number of generations that can elapse before a study is over. What was needed, Lenski realized, was an organism with a really fast life cycle, one that could zip through generations willy-nilly, allowing changes to accumulate fast enough to study their long-term evolutionary consequences.

The second unnatural—artificial—aspect of lab and agricultural studies is that selection is usually directly imposed by the investigator or breeder. Want to select for meatier cattle? Pick out the beefiest animals and let them, and only them, breed, generation after generation. This is a great means of studying the power of selection to produce evolutionary change, but it's not what goes on in nature.

Rather, out in the wild, selection is rarely so strong that only the few most phenotypically extreme individuals survive and reproduce. Instead, selection on any given trait is usually much more muted, and many selective pressures on different traits occur simultaneously, sometimes in contradictory ways. Faster mice might have an advantage, but only a small one. The fastest mice might be ten percent more likely to survive than the slowpokes, but that would mean that by chance many fast mice wouldn't make it and by luck some slower ones would. And the best-camouflaged ones might have a similar advantage, but fast mice are not necessarily well camouflaged, so selective pressures might be in conflict. The overall result is that selection is often pretty weak and probabilistic, much different from the very strong and definitive way in which lab selection studies are run.

Another way that lab selection is unnatural is that, in nature, selection is usually not consistent through time. One year, more muscular deer may have the advantage, the next year times may be tough and the lean ones survive. Indeed, as populations evolve, they may themselves alter the selective milieu—a trait that is favored when it is rare may no longer provide an advantage when it is common. Or, as populations evolve, they may alter their environment—think of beaver dams as an extreme example—and these alterations may feed back to the selective process, selecting for new traits that were not previously favored. Again, this is very different from the consistent selection pressures applied generation after generation in lab selection studies.

Lenski realized that there was a way around these problems, a strategy that had already been developed, but had not yet reached its full potential. That approach was to study evolution experimentally, in the laboratory, using microscopic organisms. Microbes have very short generations—twenty minutes or less in some—permitting plenty of opportunity for evolution on a human timescale. And instead of directing the selective process, as most previous lab studies had done, researchers could expose the organisms to a novel environment, one to which they are presumably not well adapted initially. Undoubtedly, selection would drive evolution under these conditions, but it would be the experimental environment, not the investigator, determining which ones survived and reproduced, just like in nature.

Some microbiologists had been doing this since the 1940s, but they were doing so not to study how evolution works, but rather to understand the internal workings of microbial organisms. The idea was to expose microbes to a tough environment and see what biochemical or physiological tricks they could come up with to survive. Or, in a mean-spirited but effective approach, the researchers would use molecular biology techniques to disable some bit of microbial machinery and see how the population could evolve a compensatory work-around. In this way, much was learned about how DNA works and how cells function. But for the most part, practitioners of this approach were not interested in understanding evolution; rather, they were using evolution to study how cells work. This began to change in the early 1980s, just as Lenski was finishing his graduate studies.

FAST-FORWARD SIX YEARS TO 1988. The young Lenski has now finished a postdoctoral fellowship in the University of Massachusetts laboratory of Bruce Levin, a giant in the field of microbiology and one of the few at that time using experimental studies of microbes to study evolution.* Taking up a faculty position at the University of California, Irvine, Lenski set out to establish his own research program, implementing his vision of a new approach to long-term evolution experiments using laboratory studies of the common bacterium *Escherichia coli* (*E. coli* to you and me).

Each *E. coli* individual is made up of only one tiny cell, typically about one micrometer (0.00004 inches) long, and these cells can divide

---

* There were others, and in addition to Levin, Lenski specifically acknowledges Lin Chao, Dan Dykhuizen, and Barry Hall as pioneers in the field.

as quickly as every twenty minutes when food is plentiful. The microscopic size of *E. coli* means that even a small flask can contain hundreds of millions of individuals. The more individuals in a population, the greater the number of mutations that will occur. And the more mutations a population has, the greater the likelihood that, by chance, a particularly useful one will crop up, one that will be favored by natural selection and will allow the population to become better adapted to its environment. So one could expect that *E. coli* would evolve more readily than other lab organisms with longer life spans and smaller population sizes.

Because lab scientists had been studying *E. coli* for decades, researchers knew the kind of environments in which it could live. Based on this knowledge, Lenski was able to devise an environment that they could tolerate, but that would be challenging, one in which there was plenty of room for evolutionary improvement.

As an aside, I should note that working with *E. coli* is not hazardous. True, *E. coli* is periodically in the news because of an outbreak of food poisoning and it can cause some very severe, occasionally fatal, diseases. But most types—including the strain Lenski works with—are harmless. In fact, most humans carry large populations of beneficial strains of *E. coli* in their digestive tracts, where they do important work like producing vitamin K_2 and fighting off harmful bacteria. Moreover, laboratory strains have adapted to life in a glass flask and have lost the ability to live in humans, so they definitely pose no threat. Lenski and his lab members work in typical lab frocks and don't even wear gloves, much less biohazard space suits.

On February 24, 1988, a sunny and unseasonably warm Southern California day, Lenski picked up a typical lab petri dish. *E. coli*, like other bacteria, grows asexually, each cell simply dividing into two identical daughter cells. When an *E. coli* cell is placed on the surface of a petri dish, it starts to divide, and divide, and divide, eventually producing a small mound of millions of cells, all identical descendants of that first founding cell. These mounds are called colonies. The bottom of the dish Lenski picked up was covered with a layer of goopy, translucent nutrient gelatin, with dozens of such colonies growing on its surface. All those colonies had grown from single cells of an *E. coli* lab strain called REL606.* Lenski took a small, sterilized metal needle and touched it gently to a random colony, collecting hundreds of thousands of identical cells on the tip; he then immersed the tip in ten milliliters of liquid (about a third of an ounce) in a sterile glass flask. With that, one long-term population was born. After repeating the procedure eleven times, he placed the dozen flasks†—each smaller than a teacup— into a refrigerator set to a constant temperature of 98.6 degrees Fahrenheit (just like our innards).

There was one more important ingredient in this experiment. Researchers who study *E. coli* use a wide variety of different menus to sustain their microscopic charges, some with standard biochemistry lab nutrients like powdered bits of yeast or milk proteins, others with exotic ingredients such as sheep blood or broths of pig brains and hearts. The diet fed to Lenski's *E. coli* was unusual in two respects. First, the only food present in their liquid home that they could utilize was glucose, a type of sugar used by many organisms for energy.‡ Second, unlike most lab preparations, the food was in very limited supply, so much so that every day, the population quickly increased in numbers for six hours until the glucose was depleted. At that point, the cells stopped dividing and waited quietly. The next day, a lab member would

---

* "REL" comes from Lenski's initials.

† A thirteenth flask with liquid but no *E. coli* served as a control to detect contamination. For reasons I won't get into, years later a fourteenth flask was added, thus accounting for the two rows of seven flasks I observed on my visit to the lab.

‡ Including humans. When we complain about our blood sugar being low, we are talking about the amount of glucose dissolved in our blood.

siphon 0.1 milliliters of liquid out of each flask, representing one percent of the flask's contents and thus one percent of the *E. coli* population (approximately fifty million *E. coli*), and then squirt it into a new
flask with 9.9 milliliters of fresh, glucose-infused liquid (lab scientists
use the term "medium" to refer to the nutrient-filled environments in
which their charges live). And thus the cycle would begin anew.

The *E. coli* strain used to start the experiment had been the subject
of research studies since 1918. Yet, the particular conditions of this experiment, especially the low and cyclically depleted levels of glucose,
were novel to the microbes. Presumably, this environment produced
strong selective pressures to utilize the scarce resources efficiently and
quickly. Unlike most lab selection experiments, however, Lenski was
not dictating winners and losers, sorting out which microbes would
survive to the next generation. Rather, he left the microbes to duke it
out amongst themselves, in their own way determining which constel-
lation of traits was most useful. For this reason, Lenski referred to the
project not as a selection experiment, but rather as the long-term evolution experiment—LTEE for short.

At the start of the experiment, all the individuals in any one population were genetically the same, all identical descendants from the
mother cell. Furthermore, because the different colonies in the ancestral petri dish hadn't had much time to accrue mutations, the founders
of the different populations, though drawn from different colonies,
also would have been genetically identical. What this means is that the
twelve flasks in the experiment were essentially completely homogeneous genetically—no genetic variation existed either within or between populations.* Only through time, as mutations occurred, would

---

* Actually, this is not quite accurate. Lenski introduced a mutation into half the populations just in case one population had to be distinguished from another to help detect cross-
contamination, should it occur. The mutation had no phenotypic effect and thus wasn't
affected by natural selection.

variation crop up within populations, giving the populations the ability to diverge genetically from each other.

This is how Lenski's study got around the problem faced by field evolution experiments. The environments were absolutely identical from one flask to the next—at least to the extent humanly possible. Moreover, the populations themselves started out as carbon copies, completely genetically the same. This is a real-world realization of Gould's gedankenexperiment. The tape was played twelve times simultaneously, side by side in the ice-cream chest. Same starting point, same environment. Would these simultaneous plays of the evolutionary tape lead to parallel evolutionary outcomes? Or would the randomness of mutations—one occurring in one flask, a very different one in another flask—lead evolution unpredictably in different directions? Determinism versus chance—which would prevail?

LONG-TERM RESEARCH PROGRAMS provide the opportunity for anyone to play historian. You can go back and retrace the progress of the study, noting not only how the results emerged as the study progressed, but also seeing how the interpretation of the results, the message taken from the study, changed through time. This retro-investigation is facilitated by the publish-or-perish mode of academic life. To be successful, scientists have to regularly report their results, which means that a long-term experiment will leave a rich and consistent paper trail.

Lenski's LTEE is no exception. As I sit and write this, the experiment has been ongoing for more than twenty-eight years. As the numbers taped backward on the lab's windows proclaim, sixty-four thousand generations have transpired, that milestone having been passed a few months ago. And a look at Lenski's Michigan State webpage reveals a list of seventy-five scientific publications from this project.

Let's look at the second one, published in 1994, six years after the start of the experiment and summarizing results through the first ten thousand generations. Writing in the *Proceedings of the National Academy of Sciences of the United States of America*, Lenski and Michael Travisano, who had recently completed his Ph.D. in the lab, reported that all twelve *E. coli* populations had adapted to their new environment, as gauged by the rate at which the population increased in size after the daily transfer into fresh medium. However, the extent to which the populations had become better adapted varied among the populations; at the extreme, some grew sixty percent faster than the ancestor, whereas others were only thirty percent more prolific. The *E. coli* cells were also larger than those of the ancestral population, but again the increase varied, some populations showing a fifty percent increase in cell volume, while others were as much as one hundred fifty percent bigger. Lenski and Travisano concluded that the populations had adapted in different ways, the result of the different mutations that had occurred in the different flasks. As they put it, "our experiment demonstrates the crucial role of chance events (historical accidents) in adaptive evolution."

Lenski reviewed the progress of the experiment again at the twenty-thousand-generation mark. The populations had continued to get better at living in their feast-or-famine environment—on average, populations grew seventy percent faster than the ancestral population. Differences still existed among the populations in how fast they grew, but they were referred to as "subtle variation" and deemphasized—the story line was the fact that all populations exhibited the same trend, toward a much faster rate of growth. Similarly, although no new data on cell size were reported, the parallel increase in size was emphasized and variability among populations, though mentioned, was not given much attention.

In addition, the Lenski Lab had branched out in the ways it compared the populations and had made many exciting new discoveries. All twelve populations had lost the ability to grow in flasks containing another type of sugar, D-ribose, suggesting that the biochemical machinery of the cells was changing in the same way. Several detailed comparisons of different aspects of genetics revealed that the same changes had occurred in many or all of the populations. The resulting papers concluded that not only were the populations evolving in the same way in terms of growth rate and cell anatomy, but also in underlying physiology and genetics.

In 2011, now looking back over fifty thousand generations of evolution, Lenski reaffirmed this statement, saying, "To my surprise, evolution was pretty repeatable. . . . Although the lineages certainly diverged in many details, I was struck by the parallel trajectories of their evolution, with similar changes in so many phenotypic traits and even gene sequences that we examined."

AT ABOUT THE SAME TIME that Rich Lenski was getting his start in science, almost exactly halfway around the world, another young man was taking his first college science classes. Like Lenski, Paul Rainey was fascinated by biology, studying first forestry and then botany at the University of Canterbury in New Zealand. Unlike Lenski, however, Rainey did not head off to graduate school after completing his undergraduate studies. Rather, having supported himself for several years as a part-time professional jazz musician, he immediately headed to London, the start of a year wandering through Europe, performing, exploring, working in pubs, and getting a sense of the size of the world. Returning to New Zealand, he continued to support himself playing saxophone, but finally gave in to pressure from his girlfriend's family and took a job as a sales manager for a dairy company. That only lasted three months, however, before Rainey concluded that selling milk cartons and meeting with grocery store managers was not the life he wanted. Like many young people then and now, he decided to return to school. Casting about for a suitable master's program, he found an opening for a student interested in research related to commercial mushroom production.

As part of this project, Rainey learned about a type of colony-forming bacterium called *Pseudomonas fluorescens*, which not only plays a role in mushroom reproduction, but also is beautifully irides-cent. He started to work on these bacteria, growing them in petri dishes. As the work progressed, he noticed something he didn't expect: the bacteria seemed to be changing through time in their little dishes. Some lost their coloring and became transparent, at the same time los-ing their toxicity. Others divided into multiple types within a dish.

Intrigued, he continued to dabble on the side even as his mush-room-related main work continued, trying out new growth media and environments, seeing how the bacteria adapted to different environ-ments. The master's degree turned into a Ph.D. and then he went off to England for further mushroom-related postdoctoral work, first at Cambridge and then at Oxford.

It was at Oxford that it all came together. *Pseudomonas fluorescens* is a type of bacteria that lives in soil and water. One of Rainey's jobs was to collect and examine a sample of strains of the bacteria from different host plants. One such strain was collected from the leaves of a sugar beet plant growing in the local woods.

Bringing this sample into the lab, he placed it into a glass beaker full of essential nutrients. Coming back to the lab after a weekend away, Rainey discovered that the bacteria had formed a thick, sticky mat of cells across the top of the broth. Further investigation revealed that when left undisturbed in the broth, the bacteria diverged into three different types that occupied different parts of the beaker. This was interesting enough, that a single ancestral type would give rise to three types seemingly adapted to using different parts of the environment, an adaptive radiation in miniature.

Rainey's documentation of laboratory adaptive radiation was novel, but what was really fascinating was that as Rainey reran the experiment again and again, placing the beet-loving bacteria in the broth and letting it do its thing, the same three habitat types appeared each time. Always.

When Rainey first placed the bacteria into the beaker, they were round and sleek and found throughout the broth. He called them "smooth." Fairly quickly, however, the smooth type became restricted to the interior of the broth as two other types evolved. The first of these were roundish cells with massively furrowed, crinkled edges that stick together to form mats across the top of the broth. Rainey started referring to them as "wrinkly spreaders." The other type, like the smooths, are very round, but covered with a dense, hairy fuzz, hence the name "fuzzy spreaders." Like the wrinklies, fuzzies also glom together to form a mat on the surface, but they are not as talented and quickly sink

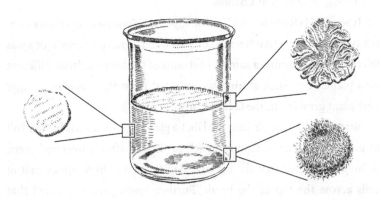

In Rainey's experiment, Pseudomonas fluorescens *diverged into three types with different shapes: smooth (left), wrinkly spreader (top), and fuzzy spreader (bottom).*

to the bottom of the beakers, periodically recolonizing the surface for a short while when viruses attack the wrinklies.

Why this adaptive diversification occurs is clear—oxygen is a limiting resource. As ancestral smooths swim through the broth, they deplete the oxygen, leading to the evolution of wrinkly and then fuzzy spreaders that can take advantage of the high oxygen levels at the broth's surface. Insect-eating lizard species adapt to different habitats to minimize competition; little cells gobbling oxygen do the same.

Lenski had demonstrated convergent evolution within a single species in the LTEE, but this was going one step further, the same adaptive radiation occurring repeatedly. Put a *Pseudomonas fluorescens* into a flask with a specific set of nutrients, leave it alone for a few days, and—voilà!—you'll get a mix of smooths, wrinklies, and fuzzies. And not only do the same three types emerge, but they do so in a predictable sequence, with the wrinklies rising to dominance first, only later followed by the fuzzies. It doesn't get much more replicated than that.

After six months at his postdoc in Oxford, Rainey—working on the side—had figured out the basics of the story. Excited, and proud of what he'd done on his own, he presented the work for the first time at his research lab's semi-annual meeting. These conclaves were held in the office of the institute's director, an eminent virologist who, like many molecular biologists of his generation, had a disdain for any biological research that wasn't entirely focused on understanding the way molecules worked.

Partway through Rainey's presentation, the director cut him off, telling him that the work was worthless because it was just documenting what happened without understanding the molecular changes that occurred as the different types specialized. Rainey was forbidden from conducting further work of this sort at the institute.

Needless to say, he was devastated. But Rainey is also, by his own description, a "stubborn bastard," and the put-down only solidified his resolution to get to the bottom of the story. The work continued, albeit furtively.

A major issue in the contingency-versus-determinism debate is the significance of happenstance, the extent to which chance events can shape the future. Of course, such events can be as important for human history as for evolution. And in Paul Rainey's story, this is where happenstance plays a role. Not too long after the lab meeting, Rainey learned of a visitor to Oxford, a researcher in residence for a year-long sabbatical. Rich Lenski.

Rainey set up an appointment to introduce himself and one meeting turned into many. Unlike Rainey's narrow-minded boss, Lenski saw the significance of the work. In addition to providing encouragement and advice, Lenski wrote a strong letter of support for Rainey to get a prestigious postdoc to continue the work. Rainey got the position in 1994 and was able to focus full-time on the *Pseudomonas* research.

A year later, Lenski chaired the renowned Gordon Research Conference on Microbial Population Biology and invited Rainey to speak. This would be the first public presentation of the work, and Rainey threw in everything he'd discovered. The talk was a tour de force, but his research program was unlike anything going on at the time, and thus foreign to the assembly of scientists. And it probably didn't help that Rainey stuck with the pet names he had coined for the three habitat specialists—"wrinkly spreader" and "fuzzy spreader" sound more like something from a children's book than part of an important research program. It probably would have been better to devise boring and technical terms—perhaps something like "Colony Growth Form IIa, Pilose" for the fuzzies. In any case, the novelty of the work, the funny names, and Rainey's own presentation style led to a lot of laughter during the presentation. To this day, Rainey is not sure how many were laughing with him and how many at him.

Nonetheless the general reception was positive and Rainey was encouraged. The next year, two years into his five-year postdoctoral fellowship, Oxford hired him on as a lecturer (the equivalent of an assistant professor in the United States). Now flush with funding, Rainey hired his own postdoctoral fellow, a young researcher whom Rainey had met at the Gordon Conference—Michael Travisano, the Lenski Lab veteran. Travisano and Rainey finished up the work, publishing their now-classic paper two years later in the journal *Nature*.

LENSKI'S LTEE AND RAINEY'S replicated microbial adaptive radiations created a new subfield of evolutionary biology. At first, the work was limited to a few labs. But that phase didn't last long. For one thing, academic generations are short—lab progeny quickly grow up, fledge, and found their own labs. By the end of the 1990s, some of Lenski's students were already faculty, continuing the long-term experimental evolution approach at other institutions. In addition, others independently took up the approach, expanding not only the researcher base, but also the diversity of organisms being studied. A quarter century after Lenski squirted *E. coli* into a dozen flasks, scores—perhaps hundreds—of labs are conducting experimental evolution studies, enough that entire conferences are now devoted to the topic.

Most of these studies have been relatively short-term—the *Pseudomonas* experiments took only ten days, for example. But increasingly researchers are emulating Lenski's approach, continuing experiments for protracted periods of time.

The LTEE is, of course, the pioneering patriarch of this type of work, and it's worth considering what it takes to keep a long-term experimental evolution study running. Every day for more than twenty-eight years, someone in the Lenski Lab has performed the transfers, moving each population from its spent environment to a newly prepared batch of glucose medium. Although the actual process takes only a few minutes, it's the organization and implementation that's impressive. The work continues day in, day out, during holidays and snowstorms, dealing with illnesses and life events. Everyone in the Lenski Lab credits long-serving lab manager Neerja Hajela as the mastermind who has made the project run so well for so long.

Only three times in twenty-eight years have the transfers not gone on according to their daily rhythm. The first time was in 1991, when the Lenski Lab moved from the University of California, Irvine, to Michigan State. Moving a lab is complicated and in this case required a hiatus in the project. Such a long break wasn't a problem. *E. coli*, like many microbes, has a superpower: it can be frozen, placed in suspended animation like an astronaut in a sci-fi movie and then later thawed out and revived, no worse for the wear. So the LTEE was simply put on hold in the deep freeze and trucked across the country. Nine months later, the populations were defrosted and the experiment picked up where it had left off.

The second and third stoppages were shorter; in the winters of 2007 and 2010, everyone in the lab was away over the holidays and so the project was put on temporary hold. This has never happened again, and the project has continued every day for the last seven years.

A GREAT VARIETY OF APPROACHES has been taken by experimental evolution researchers, both in terms of research design and the questions being addressed. Nonetheless, many of these studies have followed the Lenski-Rainey lead and established replicate populations in identical environments, probing the question of whether evolution occurs in parallel in all populations.

Given that both the LTEE and Rainey's study found that replicate populations evolved in much the same way, is there any reason to expect that other experiments wouldn't yield the same result? Put another way, if initially identical populations adapt to identical environments, why wouldn't they adapt in the same way?

Two considerations are relevant. First, to evolve, populations require genetic variation. No variation, no ability to change—natural selection, after all, works by favoring one variant over another; if there's no variation, then selection has nothing with which to work. Because the populations initially lack genetic variation, all subsequent evolution is based on mutations that occurred after the experiment begins. That, in turn, means that the evolutionary course of the populations might be shaped by which mutations occur in each replicate. Evolutionary differences among populations might arise simply because different mutations occur in different populations.

Moreover, whether a mutation becomes established in a population may depend on the order in which mutations occur—a mutation may not be beneficial if another is already established in the population, or, conversely, it may require that another mutation first become established. So even if the same mutations occur in two populations, different evolutionary outcomes could result from differences in mutational order.

The second factor that could affect whether populations evolve in parallel concerns whether there are multiple ways to solve a problem posed by the environment. As discussed in Chapter Three, species facing similar environments may not adapt convergently if they can evolve different phenotypes that produce the same functional response (good swimming abilities can result from powerful tails, forelegs, or hindlegs) or if there are multiple different functional ways of adapting to the selective circumstances (in response to a new predator, evolve long legs to better escape or camouflage to avoid detection). But we also saw that closely related populations, because of their genetic similarity, are more likely than distant relatives to evolve in the same way.

So, given these considerations, what would we expect in experimental studies of microbial evolution? In both Lenski's and Rainey's experiments, replicate populations for the most part evolved in the same way. Is this the general rule?

Evaluating this proposition is difficult because studies examine evolution in different ways. The clearest are studies that examine traits of the populations to see if they repeatedly have evolved in similar ways, such as the large cells in the *E. coli* experiments or the three different *Pseudomonas* cell types.

Another organism that has been commonly used in experiments is the familiar baking yeast, *Saccharomyces cerevisiae*, used for centuries by humans for baking, winemaking, and brewing. More recently, this yeast has taken on another role: model organism for molecular biology research. Unlike the other microbial species I've discussed, yeast are eukaryotes—like us, within each cell they have a self-enclosed nucleus that houses the DNA. This makes their biology more relevant to humans and other large organisms.

Even though they have a nucleus, each yeast individual comprises just a single cell. At least usually. A group of researchers, led by the seemingly ubiquitous Michael Travisano (now head of his own laboratory at the University of Minnesota), wanted to study the evolutionary transition from unicellularity to multicellularity, an important milestone in the evolutionary history of life. How this change occurred is a particularly interesting question to evolutionary biologists because it involves individual organisms losing their autonomy and evolving to work together for the common good. Why would initially independent cells come together to form a multicellular organism in which only some cells get to reproduce? Consider your own body—you've got cells in your brain, your eyes, your legs, throughout your body. But only a small number of your cells, those in your eggs or sperm, actually get to reproduce and pass on their DNA to the next generation. What's in it for the rest of your cells? This is a long-standing question, and Travisano wanted to get insight by studying evolution in the lab.

But how do you encourage unicellular organisms to team up? The researchers figured that by selecting for large size, they would promote the evolution of cells grouping together to form a larger mass. Early efforts failed miserably before Travisano's team hit on a winning plan. Figuring that heavier masses would sink more rapidly when placed in a liquid-filled test tube, they set up an apparatus in which cells were spun around in a centrifuge for ten seconds. Those that fell to the bottom most quickly—the bottom one percenters—were extracted and placed in a test tube to reproduce for the next twenty-four hours, at which point they were put through the spin cycle again, a process that was repeated daily for two months. This selection for rapid sinking worked just as expected, leading to size increases in all ten populations.

Exactly as the scientists hoped, the cells stuck together to form multicellular, snowflake-like conglomerates. Moreover, the mechanism

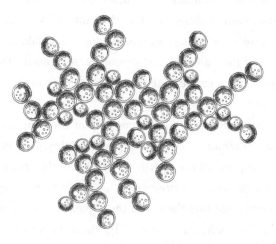

The snowflake-shaped yeast aggregations formed in Travisano's experiment

by which the amalgamation occurred was also the same in all ten populations. Rather than individual yeast cells coming together, as occurs when brewing beer, the multicellular aggregates evolved through changes in the reproductive process. Normally, yeast reproduce like *E. coli*, with one cell dividing into two cells that then go their separate ways. However, in the snowflakes, the splitting process was initiated, but not completed. One cell divided into two, but the daughter cells remained connected to each other. As a result, the structure grew as the cells continued to divide.

This experiment differs from the work by Lenski's and Rainey's groups in that the researchers directly imposed selection rather than just putting organisms into a new environment and letting nature take its course. In Lenski's parlance, this was a selection experiment, not a long-term evolution experiment. Nonetheless, the message from Travisano's studies is very much the same as that from Lenski's and Rainey's work—faced with the same selective environment, populations independently evolved in the same way.

In contrast to the work of Travisano, Lenski, and Rainey, most long-term laboratory evolution experiments do not measure phenotypic traits. The reason is simple: it's very difficult. Microbes are small and it's usually an arduous and time-consuming task to take accurate measurements of their anatomy or physiology. As a result, the extent to which phenotypes evolve in the same way is not usually evaluated in these studies.

Instead, most studies take one or both of two complementary approaches to examining evolutionary repeatability. One method is to compare the population's growth rate to that of the ancestral population. Through time, populations tend to become better adapted to new environments, and thus the size of the population—the number of individual cells—increases more quickly. Many studies have found that the increase in adaptation is very similar quantitatively from one experimental population to the next. Recall that the average increase in growth rate in Lenski's experiment was about seventy percent, with only subtle variation from one population to the next.

Another study on *E. coli* by a different team yielded similar results. The researchers established 114 populations and subjected them to very high temperatures for two thousand generations. Presumably this led to selection for physiological adaptations to withstand the rigors of living in a hot tub, but physiology itself wasn't examined. Rather, the researchers reported a very consistent increase in growth rate of about forty percent compared to the ancestral strain.

This common finding that experimental populations increase their fitness by the same amount tells us that they have become better adapted to the same extent, but it doesn't tell us how they've done it. Maybe the similarity in their increased adaptedness has come about because they've evolved the same traits, but it also could have come about through the evolution of different traits that just happen to be equally fit.

The second approach to detecting repeatable evolution is to compare the genetic changes that have occurred among experimental replicates. These days, it's possible to cheaply and quickly sequence the entire genomes of many individuals. Such studies often find that genetic changes occur primarily in the same genes across experimental populations. In the hot tub experiment, for example, mutations occurred in one particular gene in 65 of the 114 experimental populations. Moreover, even when mutations occurred in different genes, they often occurred in related genes with much the same function, a result also obtained in Lenski's studies.

A couple of caveats need to be kept in mind when considering these genetic comparisons of experimental populations. First, in almost all cases, experimental populations will acquire mutations in the same gene more often than would be expected by chance, but that's a far cry from the populations being identical in their genetic evolution. In any two of the *E. coli* hot tub populations, for example, only twenty percent of the genes that mutated in one population had also acquired a mutation in another. So, in a statistical sense, experimental populations tend to evolve genetically in similar ways, but a lot of differences evolve from one population to another as well.

Moreover, even when two populations acquire a mutation in the same gene, usually the mutations themselves are not the same, but rather represent changes in different DNA positions within the gene. A reasonable presumption would be that such mutations would produce similar phenotypic changes. However, it is always possible that different mutations in the same gene can have markedly different effects on the functioning of the gene and thus might lead to different phenotypic outcomes. Without phenotypic data, we can't be sure.

These caveats notwithstanding, it's fair to conclude that there's a lot of repeatability to microbial evolution experiments. The Lenski and Rainey experiments are the best known, but in general the others give a similar message: populations adapt at roughly the same rate and they do so—as far as we can tell—primarily by evolving similar adaptations. They tend to use the same sets of genes to accomplish these parallel outcomes. These results suggest that evolution follows the same path time and again, at least at the macroscopic level—identical populations exposed to identical selection pressures usually will evolve in very similar ways.

With one major exception.

chapter ten

Breakthrough in a Bottle

The setup of the LTEE required that the flasks be changed every day of the year, weekends and holidays included. Most lab members took part in this responsibility. Neerja Hajela, the lab manager, carefully trained newbies on the proper procedures for transferring the *E. coli* from their glucose-depleted flasks to new, freshly provisioned homes, watching them like a hawk the first couple of times they did it. Every month, Hajela put together a schedule, assigning weekend and holiday duty.

On a wintry Saturday in late January 2003, Lenski Lab member Tim Cooper was on weekend LTEE duty. It was a task that he'd performed many times before, but on a cold and blustery day with snow falling, he probably would have preferred to stay home. Nonetheless, duty called, and into the lab he went.

Cooper really enjoyed this part of the work. The experiment was already fourteen years old and had yielded significant scientific discoveries. By transferring the bacteria from their depleted homes to new, briefly resource-rich containers, he felt he was working with a piece of scientific history, one already important in the annals of evolutionary biology. Little did he expect as he pushed through the falling snow that he was about to become part of that history himself.

Once he arrived at the lab that morning, he got to work. He went to the cabinet stockpiled with sterile glassware and took out new flasks. Each one was capped with a little upside-down glass beaker, covering the hole on top to prevent any airborne bacteria from landing in the flask and contaminating it. After labeling the flasks, he grabbed a bottle of pre-made medium and carefully pipetted 9.9 milliliters into each new flask.

Now the time had come for the all-important transfers. The starved microbes had been sitting quietly for most of a day, waiting. A lucky few of them would be invited to a new banquet, where they would resume consuming, dividing, and conquering. Cooper went to the ice-cream chest incubator and removed the flasks housing the LTEE populations, placing them in a tray to carry over to the lab bench. There, a small amount (0.1 milliliter) of medium would be sucked out of the old flask and dropped into the corresponding new flask, initiating another day's worth of population growth. Cooper's routine was to remove two flasks at a time, take a quick look at them, and then place them into the tray. The self-proclaimed geek in him turned the task into a challenge, the goal being to minimize the time required to get the microbes from their old home to their new abode.

As the number of *E. coli* cells in a flask increases, the liquid becomes slightly less clear—right after each day's transfer, you could see right through it, but by the next day, the view would be slightly hazy. Cooper's task as he did his pre-transfer examination was to make sure that the liquid in each flask was appropriately murky—too clear would indicate that something had gone wrong the previous day and there were no bacteria in the flask; too opaque would suggest a bacterial population explosion, the result of contamination by some other bacterial species. This was a pro forma task—in his three years, Cooper had never encountered a day-old flask that was anything but slightly cloudy.

Cooper lifted the first two flasks out of their holders on the metal platform in the chest, making sure to keep the beakers snugly situated on top. A quick look confirmed that nothing was amiss. The second set of flasks looked the same. Next up were the flasks containing the populations known as Ara–3 and Ara+3 (it's not worth explaining what that means). When he raised these flasks, he received . . . perhaps "the shock of his life" is too melodramatic, but what he saw certainly came as a huge surprise. The liquid in Ara–3's flask was opaque and soupy. This opacity signaled a bacterial population explosion, something that wasn't supposed to happen given the limited nutrients provided to the *E. coli* each day.

Similar outbreaks had occurred several times previously in the fourteen years of the experiment, the result of screw-ups that had allowed another type of microbe to slip into the flask, one better suited to take advantage of the conditions therein. Such a misstep was a foreseeable problem, and the Lenski Lab was ready. Every day after a population was transferred to its new flask (let's call it F1), the old flask (F0) was placed in the fridge for a day for safekeeping. If on the next day, the new flask (F1) was found to be turbid and probably contaminated, the contents of that flask were discarded and that day's flask (F2) was inoculated from the old flask (F0) in the refrigerator. Essentially, the experiment skipped a day—the day that the contamination had probably occurred—and went back to the previous day's old flask to inoculate the current day's new flask. Following lab protocol, Cooper did exactly that, using Friday's refrigerated flask of Ara–3 to inoculate the new Ara–3.

Weekend duty in the Lenski Lab is a two-day affair, so Cooper returned on Sunday. To his surprise, Ara–3 was cloudy again. His curiosity piqued, Cooper withdrew a small sample from the flask and looked at it under the microscope. He expected to see a small number of *E. coli* cells and a large number of some other type of bacteria—the contaminant, whatever it might be. But the cells all looked like *E. coli*. Admittedly, most bacteria look like *E. coli* under the microscope, so this observation wasn't conclusive. Still, Cooper was excited, thinking that this might be something big.

The lab also had a protocol for persistent contamination—contamination that continued even when the previous day's flask was used. Who knew when the contaminants had slipped into the flask? Possibly they had been there for several days, but had taken time to build up their numbers to outbreak levels. The lab couldn't possibly retain every day's samples for just this eventuality—there's only so much refrigerator space. So if going back one day in time didn't cure the contamination problem, the next step was to go to plan B.

Following lab procedure, the clock was turned back a little further, in this case to a point about three weeks in the past. The reason this was possible is a result of the cryogenic capabilities of *E. coli*. The microbe's ability to survive in a state of suspended animation and later thaw out allows the experiment to be shut down for periods of time, but it also permits researchers to keep samples from previous points in the experiment, revivable if ever needed.

Every five hundred generations of the experiment (about seventy-five days), Lenski Lab members carefully transferred the unused ninety-nine percent of a flask's population to meticulously labeled glass vials and placed them in an ultra-cold freezer at −112 degrees Fahrenheit. When experimental mishaps had occurred in the past, the researchers just went to the freezer—the "frozen fossil record," as they call it—retrieved the most recent archived sample, and restarted the experiment from that point.

The freezer in which the samples are archived is named Avalon, and duplicate samples are kept in backup freezers dubbed Kyffhäuser, Valhalla, and Sheshnag. Know the significance of those names? I didn't, either. According to Zack Blount (whom we'll meet shortly), the backup freezers are "named after places in myth and legend where great heroes sleep until they are needed again."

In all previous cases of putative contamination, when the lab had restarted from the samples in Avalon, the problem had disappeared— the resurrected populations behaved normally, the flasks their normal, lucid selves.

But not this time. It took a few weeks, but the turbidity returned. Further examination confirmed that contamination was not the cause. Rather, the rightful inhabitants of the flask were on a population rampage. The *E. coli* from population Ara–3 had evolved in some way that was letting them grow to ten times their normal population size.

Ara–3's population size was much too large to be supported by the minimal amounts of glucose provided every day. Obviously, the microbes in this one population had evolved the capacity to feed on something else in their broth, something that had always been there, but which none of the other populations had ever been able to utilize. The obvious candidate was a molecule called citrate, a derivative of the citric acid that gives lemons their mouth-puckering sour bite.

In theory, citrate is suitable as an energy source for *E. coli*. In fact, when there is no oxygen, *E. coli* is able to take in citrate from the environment and feast on it. But in the presence of oxygen, *E. coli* does not eat citrate. The reason is that the job of getting citrate into an *E. coli* cell falls to a protein, referred to as a transporter, that sticks out of the cell wall, snags citrate molecules, and pulls them back inside, where they are digested. This protein is produced by a gene called *citT* that only becomes active in environments lacking oxygen. Why this particular arrangement evolved is unknown.

E. coli's inability to use citrate in the presence of oxygen is so pervasive and absolute that it is considered a diagnostic laboratory tool for determining whether a bacterium is *E. coli* or not. According to science writer Carl Zimmer, *E. coli* is "the most intensely studied species on Earth." Yet, despite the countless experiments on the organism over the past century, only a single case of laboratory *E. coli* evolving the ability to use citrate in the presence of oxygen had ever previously been reported, back in 1982.

The occurrence of citrate in the experimental medium was a fluke of history. Previous researchers had included citrate in *E. coli* experiments and since it had worked in the past, Lenski had stuck with the tried and proven recipe. Knowing of the 1982 study, he wondered whether a population might adapt to utilize it, but the idea faded as generation after generation of *E. coli* failed to crack the citrate nut.

Until the murkiness of generation 33,127. Once contamination had been ruled out, citrate utilization was the obvious next hypothesis. Initial tests were positive: when samples of Ara–3 were placed in a flask containing citrate, but no glucose, they were able to survive and grow just fine.

At this point, the job of figuring out what was going on in Ara–3 was handed to another postdoc, Christina Borland, an expert in molecular genetics with a Ph.D. from Yale. She had the challenging job of making the case airtight. First she ruled out the possibility that a contaminant, invisible to normal methods of detection and capable of eating citrate, had invaded the population. Next, she had to determine whether the citrate-munching *E. coli* were definitely Ara–3—perhaps the population had somehow been contaminated by another strain of *E. coli* that had figured out how to use citrate. Her analysis of the DNA showed that the population contained particular mutations that had long characterized Ara–3.

That left only one conclusion: this single population that had lived in flasks for fourteen years in the Lenski Lab had made a major evolutionary leap. Somehow, through the right combination of mutations and natural selection, the population had evolved an adaptation that, as far as anyone knows, this species had never been able to produce in the millions of years of its existence in the wild.* The evolutionary significance of this adaptive transformation is so great that Lenski floated the idea that the strain could be on the road to becoming a new species, a suggestion that, thirteen years on, is soon to be presented in a scientific paper. And it only occurred in one of the twelve populations. Even now, more than a dozen years and thirty thousand generations later, the ability hasn't evolved in any of the others. So much for predictability and parallel evolution!

STEPHEN JAY GOULD PROPOSED his idea of "replaying the tape" as a thought experiment, one that he figured could never be implemented. "The bad news is that we couldn't possibly perform the experiment," he wrote. But in fact, it is possible in microbial systems. The ability to freeze and revive microbes means that we can turn back the clock, replay the tape. Ice-cold samples of ancestral populations can be resuscitated and set back into evolutionary motion, and we can see whether the outcome is the same as the first time. This is a major advantage of working with microbes, one that Lenski admits he didn't fully appreciate when he initially set up the experiment. He thought he was setting up a replicated analog to Gould's metaphor—evolving twelve populations in parallel—but the ability to revive microbial populations from earlier time points means that he truly can replay the tape, go back in time, and start again.

And so in the winter of 2004, it fell to a twenty-seven-year-old graduate student named Zack Blount, newly arrived in the Lenski Lab, to hit

---

* A few populations of *E. coli* in nature are capable of using citrate in aerobic environments, but in all cases that's because they were able to swipe the necessary genes from other microbial species, rather than evolving the ability themselves.

the restart button. Blount, a soft-spoken Georgian with a surprising affinity for northern winters, didn't come to Michigan State to work with Lenski. But the fit in the original lab he hoped to join wasn't quite right, so he went looking for other opportunities. Blount liked the hypothesis-driven approach of the Lenski Lab, and Lenski saw in Blount a young man "earnest and smart, a bit quiet, motivated to do science out of curiosity and the love of knowledge."

Blount arrived in the lab at just the right moment—Borland had established that the Ara–3 *E. coli* had evolved the ability to use citrate, but this led to many more questions. Originally Blount worked on a portion of the project under Borland, but when she moved with her husband to China later that year, Lenski turned the entire project over to him. Little did either of them realize that it was going to be a decade-long job, one that earned Blount not only a Ph.D., but several years of postdoctoral research and international accolades.

Clad in what has become his trademark tie-dyed lab coat (originally green, now blue), Blount set to work. By this point, trillions of *E. coli* had lived and died in the experiment (twelve populations, thirty thousand generations, tens of millions of cells at carrying capacity every day—you do the math). For that reason, it seemed unlikely that the ability to digest citrate ("Cit+" in Lenski Lab lingo*) was the result of a single mutation: if a single genetic change could produce this ability, given the billions of mutations that must have occurred over the course of the experiment, surely Cit+ would have evolved sooner or in multiple populations.

A more likely alternative was that it took several genetic changes, occurring one after the other, for Cit+ to evolve. Distinguishing between these two possibilities—one mutation or several responsible for

---

* Blount admits that by microbiological convention, it should have been "cit+," but by the time they discovered their error, they were so enamored with the capital C that they've stuck with it.

the evolution of Cit+—is straightforward, at least in principle. Back went Blount to Avalon's frozen fossil repository, this time to see if he could get Cit+ to evolve again from an ancestral Ara–3 population that lacked the capacity to ingest citrate ("Cit–"). If the incorporation of particular prior mutations was necessary for citrate evolution, then only relatively recent populations should be able to evolve Cit+ because only those populations, and not earlier ones, possessed the predisposing mutations. By contrast, if it only took a single mutation, then the ability to become Cit+ should be equally likely to occur in any ancestral population revived from any point in the experiment.

Straightforward, however, does not mean quick. Blount pulled out samples of Ara–3 archived at twelve different time points in the experiment, from the initial ancestor in 1988 to recently frozen populations. From each time point's sample, he isolated cells and used them to create six replicate populations, making seventy-two in all, and let them evolve for two and a half years. Four of the seventy-two populations evolved the ability to digest citrate, all four from relatively recent ancestors. These results suggested that only recent Ara–3 populations were capable of evolving to digest citrate.

Blount got bored waiting for the results from this experiment, so while it was going on, he tried another technique, one more sensitive to the evolution of slight Cit+ ability. In this approach, cells were again taken from population samples frozen at different time points and grown up to produce multiple populations of more than ten billion cells each. These populations were then placed in petri dishes in which citrate was the only thing to eat for up to three weeks. Under these conditions, only the rare Cit+ mutant would be able to grow to form a colony. Out of 3,200 populations that Blount examined, only thirteen, about a third of one percent, evolved to become Cit+. Again, most of these were very recent, the earliest coming from the ancestral population frozen at twenty thousand generations.

Two results are clear from these replays. First, the evolution of Cit+ occurs only rarely, even among populations maintained under identical conditions. Second, the ability to digest citrate is not conferred by a single mutation, but rather by several, all of which apparently occur quite rarely. Something must have evolved in the population shortly before the twenty-thousand-generation mark that set the stage for the subsequent evolution of Cit+ capability. This requisite combination of very rare events explains why it took more than thirty thousand generations for the ability to arise in one population in the LTEE experiment and why it has never arisen again in the other eleven.

The description of Zack Blount's project doesn't do justice to the amount of work it took to get these results. He's become famous in evolutionary biology circles for photographs of himself sitting in the lotus position—legs crossed, eyes closed, forefinger and thumbs making a circle in meditation mudra—wearing his colorful robe in front of a gigantic tower of stacked petri dishes, composed of as many as thirteen thousand petri dishes that had been used in the course of a single one of these experiments.

FIVE YEARS AND FORTY TRILLION *E. coli* cells elapsed between when Blount started working on the project and when the first paper was published announcing these results. By the time it came out, in the summer of 2008, he was already on to the next stage of the project, figuring out exactly what the mutations were.

One advantage of a long-term experiment is that new technological capabilities emerge, making possible in the later stages what could only be imagined early on. In this case, that advance was the ability to easily and cheaply sequence the genome of entire organisms. When genome sequencing first became possible in the last years of the twentieth century, obtaining the entire sequence of an organism would cost millions of dollars and take years. But by 2008, the price was $7,000 and the wait only a month.* Armed with this capability, Blount and lab colleagues set out to sequence Cit+ and Cit– *E. coli* to identify the genetic changes responsible for the evolution of citrate digestion.

I won't go into the biotechnological details of how it was done, but it took another four years, during which time Blount completed his Ph.D. and stayed on in Lenski's lab as a postdoctoral fellow. And working his molecular magic, he was able to figure out what happened.

To do so, Blount sequenced the genomes of twenty-nine *E. coli* cells from populations throughout the experiment's history. All Cit+ populations shared a mutation not seen in any Cit– populations. As you will recall, *E. coli* naturally can capture citrate in the absence of oxygen by turning on the *citT* gene, which causes the cell to produce transporter proteins that poke out of the cell's membrane and latch on to nearby citrate molecules. What happened in the Cit+ *E. coli* cells is that a duplicate copy of that gene was made. This happens all the time in most organisms—when new cells are produced, DNA copies itself, and sometimes a mistake in copying leads to the production of two copies of a gene, one attached to the end of another.

Normally, the *citT* gene, which produces the citrate-snagging transporter protein, is activated when oxygen levels are low. In contrast, *rnk,* a gene that occurs close to *citT* on the chromosome, turns on when oxygen levels are high, rather than low. Just by chance, when the second copy of the *citT* gene was accidentally created, it ended up being placed right next to the activation switch for the *rnk* gene. This rewired the *citT* copy to be turned on along with *rnk* in the presence of oxygen. This happenstance of molecular miscopying in the DNA replication process gave Cit+ *E. coli* the ability to ingest citrate in the presence of oxygen.

---

* By 2013, the price had dropped more than ninety-nine percent from the 2007 level. The quality was vastly improved as well.

As a result of Blount's work, we now have a good idea of why Cit+, despite being so beneficial, has evolved so rarely in the LTEE populations. Several highly improbable events had to happen. The key genetic change permitting utilization of citrate was a duplication event, in which part of an entire gene was copied into the genome an extra time. Moreover, not only did *citT* have to be copied, but the copy had to land in just the right place for it to be activated in an aerobic environment. Blount's replay experiments show that in the right genetic context such mutations do occur, but they do so very rarely.

This rarity, however, does not fully emphasize the improbability of Cit+ evolution. Not only does the *citT* gene have to replicate itself and the copy have to land in the right place, but such duplication only leads to Cit+ evolution when it occurs in already prepared populations. Something else had to evolve first to make Cit+ possible, and this something hadn't arisen in the Ara–3 flask in the first twenty thousand generations.

Finding the "potentiating mutation"* or mutations—that is, the mutations that make subsequent Cit+ evolution possible—is a challenging task. After another three years, a team of Blount's collaborators at the University of Texas at Austin finally identified one potentiating mutation and figured out why it was initially favored by natural selection (remember that natural selection has no foresight; it will not favor a mutation because it will be useful in the future. Consequently, the potentiating mutation either must have had some benefit unrelated to citrate utilization or it evolved by chance, which is unlikely in large populations). Even now the work is not entirely finished—it seems there is a second mutation involved in potentiating the evolution of Cit+, still under study in Texas.

There's still more to the story. In addition to potentiation and the

---

* A term coined by Blount.

mutation that conferred the initial ability to ingest Cit+, there was a third phase in the evolution of Cit+ capability. Once the gene duplication occurred, Ara−3 could utilize citrate, but only very poorly. The mutation didn't provide much of an advantage and thus it wasn't strongly favored by natural selection. Indeed, for more than fifteen hundred generations, the mutation was present in the population, but at a low frequency. It was only when yet another mutational event occurred—the further duplication of the Cit+ mutation so that individuals had multiple copies of the aerobically activated gene—that citrate utilization improved enough for the trait to rapidly spread through the population. Blount's analyses indicate that this third event was not as rare as the first two, but it still took a while before it actually happened in the population.

The Cit+ saga has important implications for understanding the evolutionary process. On one hand, it illustrates how major evolutionary advances occur. Usually, any complex trait—an eye, a kidney—is not the result of a single mutation that creates a new structure from scratch. Rather, such innovations usually occur in stages, through several sequential steps.

Moreover, the Cit+ story shows that evolution is not necessarily predictable, that replaying the tape may not necessarily lead to the same outcome—not only did Cit+ evolve in only one of Lenski's twelve populations, but it did so in an extremely small percentage of Blount's reruns. The conclusion is clear: a set of mutations, occurring in just the right order, can have a major impact, sending evolution down a different, unrepeated path.

"THE TENSION BETWEEN CHANCE versus necessity." That's how Rich Lenski describes the key question underlying his thirty-year research program with *E. coli*. Indeed, the LTEE has been remarkably successful at addressing this issue, producing one fascinating discovery after another. But its impact has been much greater than the scientific fruits of this one research project. The LTEE, with help from other early pioneers, has spawned an entire industry of researchers conducting similar experiments in which replicate populations, established with identical founders, are allowed to diversify under the same conditions.

Although many experiments are currently under way, enough have been completed that the broad outlines are apparent. And the situation is much more nuanced than Lenski would have predicted back in 1988—he definitely came from the chance camp and expected outcomes like Cit+ to be the rule. Quite the contrary, necessity usually triumphs. Experimental populations facing the same natural selection pressures usually adapt in the same way, producing a roughly similar increase in fitness.

Despite this similarity, a closer look reveals that the evolutionary paths are not identical. Often, the adaptations are similar, but varied. Wrinkly spreaders from different populations differ in the fine details of shape and structure from one population to another; yeast snowflakes, though all serving to increase the rate at which they settle to the bottom, can be substantially different in size and configuration.

There are differences at the genetic level, too. The phenotypic similarities in different populations usually result from mutations in the same genes. But not always—sometimes mutations in different genes produce a similar phenotypic outcome. Recent work in Paul Rainey's lab, for example, has shown that there are sixteen different genetic pathways that can produce the wrinkly spreader phenotype. And even when the mutations are in the same genes across populations, the exact changes are rarely the same. As we'll see in Chapter Eleven, this genetic indeterminacy underlying phenotypic convergence can have important implications for subsequent evolution. Still, the usual headline from these studies is the extent of parallel evolution. Necessity trumps chance. Usually.

Radically different evolutionary responses in these experiments are rare, but the Cit+ story is not the only instance. In another *E. coli* study, a different research group found that half of their *E. coli* populations diverged into two types, with one that used acctate more efficiently than the other. A similar divergence has also occurred in one of the LTEE populations. Another example comes from a study in which some, but not most, experimental virus populations evolved an entirely new way to attack their *E. coli* prey. These examples are striking demonstrations that evolutionary replays, with identical conditions from one iteration to the next, can still yield different outcomes. It's not common, but it does happen.

The genius of conducting evolutionary experiments on microbes is the amount of evolution that can be packed into a short period of time, at least as measured from a human perspective. A sixty-thousand-generation experiment on fruit flies would take a thousand years, and for mice, ten times longer.

Yet these experiments may not be long-term enough. Sixty thousand generations is a geological blink of the eye. Species persist for millions of years. Maybe these experiments, though vastly longer and more informative than anything that has come before them, are still not long enough. Rich Lenski acknowledges the possibility.

What might we find if Lenski's academic descendants keep the LTEE going for decades more? Surely, the longer an experiment runs, the greater the chance that the rare, unlikely but beneficial combination of mutations will crop up, leading to more Cit+ cases. Maybe three hundred years from now, in generation six hundred thousand, all twelve populations in the LTEE will have become Cit+. What seems unpredictable in the shorter term may be inevitable in the longer term.

In 2002, before population Ara–3 had gone its own citrate-munching way, the LTEE seemed to be a strong confirmation of the predictability of evolution, a powerful counterargument to Gould's vaunted contingency. Now, however, the take-home message from these studies—fascinating as they are—is not so clear. Do the eleven populations marching in Cit– lockstep demonstrate that evolution is usually repeatable? Or does Ara–3 show that Gould was right, that evolution is unpredictable? The answer is: both.

In an interview published just as I was wrapping up this book, Lenski tacked back toward his initial views from three decades ago. After noting the many parallel changes that have occurred in the LTEE Twelve, he also noted the differences, not only Cit+, but several other changes that have cropped up in only one or some of the populations. "Against this backdrop of parallelism, or repeatability, the longer the LTEE has been running, the more we see that each population really is following its own path," he said, "and both sets of forces—the random and the predictable, as it were—together give rise to what we call history."

chapter eleven

Jots, Tittles, and Drunken Fruit Flies

L et's recall again what Gould said: "I call this experiment 're-playing life's tape.' You press the rewind button and, making sure you thoroughly erase everything that actually happened, go back to any time and place in the past. . . . Then let the tape run again and see if the repetition looks at all like the original."

This is precisely what Zack Blount did. Through the magic of microbial biology and the Lenski Lab's deep freezers, Blount was able to rewind the tape, to re-create the conditions that had existed in the past, and then let evolution run its course again.

But is that really what Gould had in mind? After all, the title of Gould's book alludes to the key scene in *It's a Wonderful Life* in which George Bailey's guardian angel shows him how life in Bedford Falls would have been very different if George had never existed.

Angel Odbody's ploy is not simply rewinding the tape to an earlier time and hitting the play button. Rather, it amounts to rewinding the tape, but changing one key feature: the presence of George Bailey. As a result, the *It's a Wonderful Life* story is not the equivalent of Blount's experiment. Clarence Odbody didn't say, "Let's go back in time, start over with everything the same, and see if the town's history unfolded in the same way." Rather, he asked, "Would the history have been different if the circumstances had been slightly different, specifically if you hadn't been there?"

Gould summarized the lesson from *It's a Wonderful Life* in a way that differed from his previous description of replay experiments: "This magnificent ten-minute scene is both a highlight of cinematic history and the finest illustration that I have ever encountered for the basic principle of contingency—a replay of the tape yielding an entirely different but equally sensible outcome; small and apparently insignificant changes, George's absence among others, lead to cascades of accumulating difference." Applying the lesson to evolution, Gould added an important proviso to his earlier scenario: "any replay, *altered by an apparently insignificant jot or tittle at the outset* [my italics], would have yielded an equally sensible and resolvable outcome of entirely different form."

Blount described his research as a direct implementation of what Gould had suggested. More generally, the entire LTEE project has been portrayed as directly analogous, the only difference being that the replays occurred simultaneously in multiple flasks, rather than sequentially through time. Were the Lenskiites mistaken in their claimed fidelity to Gould?*

JOHN BEATTY IS THE NICEST GUY you could ever meet. He even looks warm and friendly, a bit avuncular, with a peppered white mustache, strong chin, and slightly receding hairline, dressed in a worn leather jacket or a cardigan, comfortable in his own skin. A native Texan—

---

* The LTEE's fealty to Gould was made evident in a footnote to Blount's first paper, in which the authors noted that the Ara–3 replay experiment was begun on the third anniversary of Gould's death and terminated on the sixty-sixth anniversary of his birth.

perhaps that's the reason—Beatty is now a professor of the philosophy of science at the University of British Columbia in Vancouver.

Philosophers of science are interested in how science works. Not the specific knowledge learned about a particular lizard or neutrino, but the process of science—how scientists go about studying natural phenomena, come up with ideas, test their hypotheses, reject some, refine others.

Evolutionary biology is a particular challenge to philosophers of science. It does not fit the standard notion of how science works—itself a caricature—in which a crucial experiment decisively settles the question. Rather, evolutionary biology involves history, figuring out what happened in the past, asking questions not amenable to the experimental method (what experiment can explain the evolution of a giraffe?). I've already discussed how studying evolution can be similar to a detective story, a whodunit whose methods share as much with the study of history as they do with other sciences. This is one of Beatty's many interests, the distinction between history and science and "the respects in which evolutionary biology is as much like the former as it is like the latter," as he says on his website.

A related and long-standing interest of his concerns the role of chance in evolutionary biology. So it was natural that when Gould published *Wonderful Life*, with its emphasis on the role of history and contingency in evolution, Beatty took notice.

As the years went by and scientists began to examine Gould's ideas, Beatty went back and reread *Wonderful Life*. And reread it again. And he realized something that everyone else had missed. Seventeen years after Gould's book appeared, Beatty published a paper pointing out that Gould was confused about what he meant by the term "contingency."

Beatty recognized that the word has two different meanings in common usage. The first is unpredictability—"we have to be prepared for all contingencies." In the Gouldian tape-replay sense, what does unpredictability mean? It's not that something unpredictable in the environment happens—like a flood or a lightning strike—that causes evolution to occur in a different way. That doesn't count, because the premise of the rewind metaphor is that everything is the same, not only the environment, but the same external pushes and prods.

If the environment is the same in the replay, where does the unpredictability come in? Beatty points out the obvious possibility: differences in the mutations that occur. Biologists generally consider mutations to be unpredictable. We know that some parts of the genome experience mutations more than others, and that certain circumstances—such as exposure to cosmic rays or some chemicals—can affect the mutation rate. But we can't predict which DNA site will experience a mutation, much less what that mutation will be. For all intents and purposes, it's reasonable to treat mutation as an unpredictable, random occurrence.* As a result, we would expect replay populations to experience different mutation histories.

The question is whether such unpredictability could lead to evolutionary indeterminism. Evolution requires genetic variation and, thus, populations with different variation can evolve in different ways. Consider a population in which all individuals have blue eyes. At this point, the population can't evolve a different eye color—none of the individuals have genetic variation for any other color. But if in one population a mutation occurs that produces brown color, then it's possible

---

* Some might suggest that if we rewound the tape to the exact same point, everything being completely identical, then the history of mutations would be identical as well. That is, there must be some physical or chemical cause of a mutation; consequently, if circumstances are truly completely the same, then the outcome would have to be the same. Such a view, with an intellectual history that traces back to the French mathematician Simon-Pierre Laplace, would mean that, by definition, replaying the tape would lead to the same outcome. But this perspective ignores the fact that there is true indeterminacy at the level of quantum mechanics and hence it is at least possible that indeterminacy could resonate from the subatomic to the molecular level. In any case, for the purposes of this discussion, I will consider mutations to be unpredictable, a potential source of non-parallel evolution in replicate populations.

for that population to evolve to be brown-eyed. Yet, in another population, the brown-eye mutation might not occur, but a mutation for green eyes might, allowing that population to evolve in a different way. If mutations are unpredictable and the occurrence of particular mutations affects the direction in which evolution heads, then an evolutionary rewind might lead to a different outcome.

That is precisely the hypothesis that the LTEE addresses. And the answer, at least in this case, is clear: to a substantial extent, evolution is predictable, even if the history of mutations is unpredictable. Start with identical circumstances and you'll usually—but definitely not always!—get a pretty similar outcome.

BUT "UNPREDICTABILITY" IS only one sense of contingency. As Beatty realized, a second meaning exists. This second definition refers to what is termed "causal dependence," which results when the occurrence of an event is predicated on something else happening first—the occurrence of event B is contingent upon the occurrence of event A. Your existence is the outcome of a series of events, starting with your parents meeting, through stages of their courtship, right up to the point of their coupling at the specific time that led to your conception. Change any of those events and you wouldn't be here. Someone else—the result perhaps of a different sperm of your father's fertilizing the same egg of your mom's—might be instead, but that person wouldn't be you. Your existence is contingent upon all those events happening just as they did.

Gould put it eloquently in *Wonderful Life*:

> Historical explanations take the form of narrative: E, the phenomenon to be explained, arose because D came before, preceded by C, B, and A. If any of these earlier stages had not occurred, or had transpired in a different way, then E would not exist (or would

be present in a substantially altered form, E', requiring a different explanation). Thus, E makes sense and can be explained rigorously as the outcome of A through D.

I am not speaking of randomness . . . , but of the central principle of all history—*contingency*. A historical explanation [rests] . . . on an unpredictable sequence of antecedent states, where any major change in any step of the sequence would have altered the final result. This final result is therefore dependent, or contingent, upon everything that came before—the unerasable and determining signature of history.

This is what Gould meant by "jots and tittles." Change B by a jot, and E doesn't result. Tittle C, and you don't get E.

The differences between these two meanings of "contingency" might seem to be a matter of mere semantics. But Beatty suggested that they were much more, that the different definitions had important implications for how we view evolutionary determinism. On the one hand, the unpredictability view of contingency suggests that evolution is inherently indeterminate—start from the exact same circumstances, experience the same environmental changes, and yet the outcome might still be different. On the other hand, the causal-dependence view looks not at the beginning, but at the end. Determinists such as Conway Morris would suggest that the outcome is foreordained, that there are a few adaptive solutions that will repeatedly evolve regardless of where a population starts and what happens along the way. The Gouldian retort is "not so": the end result is critically dependent—contingent—on the particular circumstances that come before it.

JUST AS LENSKI'S LTEE catalyzed the field of experimental evolution and led to a huge enterprise of like-minded research, Beatty's paper has had a similar effect on philosophers of science. A cottage industry of philosophical work has emerged in the subsequent decade, debating the semantic nuances of the word "contingency" and divining ever more nuanced—and in some cases, far-fetched—explanations of what Gould had in mind.

Nonetheless, Gould's ambiguity is important because its implications extend beyond university philosophy departments. In particular, Gould claimed that replaying the tape of life was only a gedankenexperiment—a "thought experiment"—but microbial evolutionists proved otherwise: the LTEE and much of the research that followed were explicitly devised as a way to conduct the evolutionary replay experiments that Gould thought were impossible. Yet, as Beatty demonstrated, Gould actually conflated different ideas about contingency and determinism under his replay rubric. And as Beatty also showed, different researchers have used different meanings of contingency and thus constructed fundamentally different research programs.

The studies I discussed in the previous two chapters all follow the general setup of the LTEE—start with identical populations, place them in identical environments, and study the extent to which they follow identical evolutionary paths. This is clearly a test of the unpredictability definition of contingency, taking Gould literally when he says to replay the tape from the same starting condition, with all populations experiencing the same environmental conditions generation after generation to see if the outcome is predictably the same.

But what about the causal-dependence notion of contingency, the idea that evolutionary outcomes are critically dependent on the particular course of history? Gould's prescription here is clear: "Alter any early event, ever so slightly and without apparent importance at the time, and evolution cascades into a radically different channel. This . . . represents no more nor less than the essence of history. Its name is contingency."

This is where Gould's insignificant jots and tittles come in. He's saying that we don't just go back to some point in the past and start again from the same conditions. Rather, we go back, but change something, either in the starting conditions or something that happens along the way. As one biologist remarked, Gould's idea could be restated as "Go back in time five hundred million years, move one trilobite two feet to the left, and see if evolution unfolds in the same way."

How one would devise such an experiment seems straightforward. Simply set up a bunch of populations in identical environments and then subject them to a variety of different jots or tittles and see if they still evolve in parallel.

What form would the jots and tittles take? Let's take the LTEE as an example—what could a researcher do to test the resilience of evolutionary outcome to altered circumstances? Here are some ideas that occurred to me (remember, of course, that not all populations experience the same perturbations; the point of these experiments is to test whether the perturbation alters the course of evolution compared to a population not experiencing it): leave a flask at room temperature for a month instead of putting it back in the incubator; inoculate a flask with 0.001 milliliters of medium rather than the normal 0.1 milliliters; put a flask in an incubator that had a light on inside; put three times the normal ration of glucose into the broth for two days; put pink dye in the broth. These are just off the top of my head and from someone naïve about microbial science; surely, microbiologists could come up with a much more interesting set of experimental perturbations.

I am unaware of any experimental study of this sort and it's easy to understand why. These studies take a lot of effort to set up and run. And the perturbations being discussed are, after all, pretty minor. It seems most likely that they won't have any long-lasting repercussions. Consequently, this seems like an experiment with a low probability of getting an exciting result and a high probability of getting the expected outcome. Results of this sort usually don't attract much attention and can even be hard to publish. For this reason, studies like this can be unattractive, particularly to young scientists who need publications to advance their careers.

ALTHOUGH NO ONE HAS EXPLICITLY TESTED what we might call Gould's "resilience to altered conditions" hypothesis, some researchers have gone halfway by starting with populations that are genetically differentiated from each other. Why they are different we don't know—we have no record of their different jots and tittles—but regardless, they have experienced different histories, and these studies have asked whether their historically derived differences affect future evolution. Or, put another way, would genetically different populations facing the same environmental conditions evolve in the same way?

As an extreme example, suppose there were two populations of dogs, one made up of small pooches like schnauzers and chihuahuas, the other with large dogs like greyhounds and German shepherds. Suppose, further, that they occurred in a place where a new type of large predator, say, a tiger, arrived (perhaps they were on an island and the tiger colonized it from the mainland). The two canine populations might adapt to this predation pressure in different ways, the small dogs adapting to become camouflaged and inconspicuous, the larger dogs evolving longer legs to escape by running away more quickly. Certainly, it's not hard to imagine that the different genetic constitutions of the two populations would predispose them to adapt in different ways to the new predation threat.

The effect of genetic differences on the evolution of populations exposed to the same selection pressure was first investigated in a laboratory experiment on fruit flies in the mid-1980s. As anyone who's let a banana get a little too ripe knows, fruit flies congregate around rotting fruit. And one of the things that a fruit does as it rots is ferment, producing alcohol. As a result, the flies live in an environment thick with alcohol fumes, like spending your life in a brewery. And what happens if a fly overdoes it, soaking up too much alcohol? It gets drunk, just like you and me (well, at least like me): at first it runs around excitedly, bumping into things. Then it stumbles, it staggers, and it falls down. Eventually it falls over and doesn't get up. And the hangover is no better. The fly gets up. It falls down. It takes things slow. Probably, it swears off breathing alcohol for a while, until the next overripe banana proves too enticing.

We humans differ in our susceptibility to alcohol, and at least some of this difference is genetically based. Assuming that the same was true of fruit flies, Fred Cohan, then a postdoc at the University of California's Davis campus (now a professor at Wesleyan University), set out to ask whether the flies could evolve to become more tolerant of alcohol. More to the point, he wanted to know whether populations from different places would evolve in the same way or whether the genetic differences among the populations, evolved for whatever reason, would lead them to adapt in different ways.

As a graduate student at Harvard, Cohan had studied the population biology of fruit flies, examining the extent to which populations of the same species differed genetically. Working at a New England university, Cohan had been confined to the laboratory, working on flies sent to him by others.

But moving to California opened up new vistas. If a guy's going to spend his life staring at fruit flies in little vials in the lab, the least he could do is go outside and collect the flies himself. Especially on the West Coast, where collecting flies requires road trips to scenic areas. And doubly so when he's just married a special ed teacher who not only puts up with his entomological eccentricities, but enjoys fly-collecting expeditions.

So shortly after arriving in Davis in the summer of 1982, the Cohans hopped into a university vehicle and started toodling north, winding their way through Oregon and well into Washington state. Their goal was to collect fruit flies at localities spanning the West Coast. Eventually, the collections would be used in a selection experiment to see if geographically differentiated populations adapted in the same way to the same selective pressure. But first, the Cohans had to catch their flies.

If you wanted to capture fruit flies, where would you go? Fruit flies like fermenting fruit, so you'd need to find a place where fruit is rotting. I've heard of scientists collecting their specimens from dumpsters behind fast-food restaurants, but Cohan had a better idea—he'd go to the source of the fruit itself. So he and his wife meandered along back roads, looking for fruit farms. Remember, this was before the internet era. They couldn't go online and search Google Maps for the nearest farm. Instead they drove around in likely areas until they came across an orchard.

Farmers were surprisingly receptive to the idea of a young couple driving up and asking for permission to wander around the farm chasing flies; in fact, they were intrigued that the annoying little pests actually had some value, that they could help us learn something of possible importance. And Cohan was pleased to be able to tell them that the fruit flies weren't actually doing any harm.

So how do you catch a fruit fly? When I first thought about this, I pictured running around with a butterfly net, zigging and zagging, swooping down to snag a fly in midair. I've done that with butterflies and it's loads of fun. But not for fruit flies. Instead, what you do is set out a bucket full of fruit fly delectables and wait for dusk, the fruit fly witching hour. It's kind of like putting out shrimp and Cajun sauce at a five-thirty cocktail party, except in this case the appetizer is a crème de banane of sorts, made by mixing ripe bananas with live cooking yeast and grape juice and letting the yeast grow until the odor is just right.

Fruit flies can't resist it. Then all you have to do is put a net over the bucket, give the bucket a tap to scare the flies up into the net, twist your wrist to close the net, and the flies are enmeshed. In no time at all, you've got vials of fruit flies.

Over the course of two field trips, the Cohans visited a multitude of farms and orchards, eventually setting up lab populations from nine sites spanning the West Coast from San Diego to Vancouver (the Canadian samples having been contributed by a colleague). Back in Davis, Cohan subjected each population to selection for alcohol tolerance, allowing only the most alcohol-insensitive flies to breed every generation.

This sounds pretty straightforward, but how does one measure the effect of alcohol on a fruit fly? In humans, you could give a bunch of people a couple of stiff drinks and then measure their ability to walk a straight line, speak coherently, and so on, just like a cop does to someone pulled over for suspected DWI. You could do the same for fruit flies (sans the speech test), exposing them to a certain amount of alcohol fumes and seeing how they respond. But the problem is that this approach would be very tedious. A selection experiment of this sort usually involves hundreds, if not thousands, of individual flies per population. Watching that many fruit flies for a few minutes each to collect the data would take an enormous amount of time.

Fortunately, Ken Weber, an enterprising Harvard graduate student and a friend of Cohan's, came up with a solution—the inebriometer! It works like this: release a thousand fruit flies into a four-foot-long glass tube, sealed at the top. Flies like to hang out high up, so they move upward, flying and crawling around. At the top of the tube, insert a rubber hose that steadily pipes in alcohol fumes that exit through another hose at the bottom. As time passes, the flies get tipsy, but some more so than others. When they become inebriated, they lose the ability to fly and start falling. Spaced out alongside the inside of the tube are a series

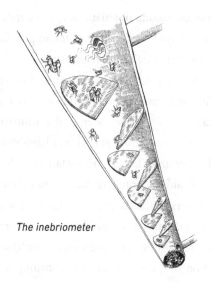

The inebriometer

of sloping shelves that give a tumbling fly a chance to regain its footing. As a result, the slightly tipsy flies might momentarily start falling, but usually are able to catch themselves on one of the shelves. But once the flies become really drunk, they are so incapacitated that they roll down off one shelf after another, eventually falling all the way to the bottom of the tube, where they land on a screen and can be removed. Eventually, only the most alcohol-tolerant flies remain, and they're the lucky winners, allowed to mate with each other to produce the next generation.

At the start of the experiment, variation existed in alcohol tolerance—some flies were stone-cold drunk and at the bottom of the tube within a few seconds, while others were still flying after a half hour. The average fly passed out after about twelve minutes, and flies from northern populations were able to last a little bit longer than the southerners.

Twenty-four generations later, all populations had evolved substantially greater alcohol tolerance. But this increase was not the same across all populations: flies from British Columbia buzzed about for almost fifty minutes on average before blacking out, whereas the Southern California flies were lucky to make it to the forty-minute mark. In other words, subjected to the same selective factors, the northern populations adapted to a much greater extent than the southern populations.

What had started as slight variation among populations led to a much greater difference as a result of selection. Genetically different populations had responded differently to the same selective pressure.

A similar experiment was recently conducted on yeast. The researchers took strains of budding yeast from six very different environments, including oak trees, cactus fruit, ginger beer, and a woman's vagina. Three samples of each strain were then placed into laboratory vials containing glucose as a food source (lab experimentalists do like to feed their charges glucose!). Would the populations, having evolved in very different ways in their diverse habitats, adapt in the same way to their new glucose diet? Five months and three hundred yeast generations later, the researchers measured a suite of traits on each population including growth rate, population size, cell size, glucose consumption rate, and the rate at which ingested glucose was converted into more yeast cells. Despite substantial evolutionary change in all traits, the populations continued to exhibit considerable variation; not only had they not converged on values for the traits, but in some cases they had become more dissimilar as they adapted to the common environment.

The key difference between these two studies and the ones discussed previously is that in the LTEE-like studies from Chapter Ten, the populations initially were identical. In contrast, in these studies the experimental populations started out different; having evolved separately for some unknown period of time, the source populations had accumulated genetic and phenotypic differences. And the result was clear. When they're identical initially, populations generally respond to selection in the same way; when they start out different, evolutionary responses can be quite divergent. Score one for Gould: change the conditions at the outset and a different evolutionary course may result.

But these results are a little unsatisfying because we don't know what happened to cause the populations to diverge in the first place.

What constitutes the jot or tittle that sets the stage for later divergence? One obvious possibility is that some sort of natural selection pressure affects one population and not another. The evolutionary response to such selection leads to genetic change. This altered gene pool then can affect subsequent evolutionary direction.

Testing this more specific scenario in the lab is straightforward if time-consuming. Expose initially identical populations to different environments over many generations. Then once they have adapted to the different environments, expose them all to the same new selective environment and see if they adapt similarly or whether their evolved differences cause them to adapt in different ways.

Surprisingly few studies of this sort have been conducted and the results have been mixed. In some studies, the populations, despite their initial differences, have evolved to become very similar, erasing their pre-existing differences. But in other experiments, populations have not converged despite experiencing the same environment. In other words, they exhibit the hallmark of contingency: what happened in the past affects what will happen in the future. It is not hard to extrapolate from studies like this to Gould's jots and tittles—evolutionary adaptation in response to past events can affect subsequent evolutionary trajectories.

But two populations don't have to be subjected to different environmental conditions to diverge genetically. Even populations experiencing similar selective pressures may not adapt in exactly the same way. As discussed in Chapter Ten, microbial evolution experiments indicate that even though genetic changes are often quite similar—often involving the same gene—the exact changes at the DNA level usually differ from one population to the next. Is it possible that such tiny genetic differences could predispose populations to evolve differently in the future?

———

TWO THOUSAND GENERATIONS into the LTEE, the twelve populations had increased their fitness about the same amount. Reporting this result in the first paper ever published on the experiment, Lenski had suggested that the populations were all evolving in the same way. Nonetheless, he recognized that another explanation was possible, that the populations were finding different ways to adapt to their new conditions, and the rate of adaptation just happened to be about the same in all of them.

These two possibilities led to different predictions about the genetics of the populations. The parallel adaptation hypothesis suggests that the genetic changes that had occurred in the populations were pretty much the same, whereas the disparate-adaptations-with-comparable-effects-on-adaptedness hypothesis suggests that the populations had experienced very different genetic changes. But this was back in the early 1990s, when investigating genes and genomes was still mostly a pipe dream. How to distinguish these two possibilities was not obvious.

The job of solving this conundrum fell to none other than Michael Travisano. Before snowflake yeast, before wrinkly spreaders, Travisano's research career began in the Lenski Lab (I've already mentioned one paper from his Ph.D. work in Chapter Nine). Travisano initially came to the lab as a technician with a background in cell biology, having done research on hamster ovary cells and what causes them to become cancerous. In retrospect, he realizes that he was doing experimental evolution studies, but they weren't portrayed that way. Rather, they were trying to figure out what causes a cell to become metastatic, looking for repeatable responses to particular experimental treatments.

With that background and the LTEE just under way, Travisano

pondered how to assess the extent to which evolution was repeatable. Working with Lenski, he devised an elegant experiment to figure out whether the twelve LTEE populations were all adapting in the same way. The trick, they realized, was to place the populations in another environment and see how they fared. If the populations had all evolved in the same way genetically as they adapted to the LTEE conditions, then—being genetically similar—they should all fare equally well in the new environment. If the populations had instead evolved different genetic adaptations to LTEE conditions, then they might vary in how well adapted they were to the new conditions.

To test this idea, Travisano took samples of the twelve *E. coli* populations from generation 2,000 and placed them in a different environment. Instead of providing them with glucose as their source of energy, he provisioned the medium with a different type of sugar: maltose.

Over the first two thousand generations of the LTEE, all of the populations had become much more proficient in using glucose and thus grew substantially faster than their ancestor had when given that food. How would this adaptation to glucose affect their ability to use maltose? To compare them to their original state, Travisano went deep into the frozen archives, resuscitated the ancestral LTEE population, and measured how well it grew on maltose.

On average, the ability to use maltose had not changed at all. But that average concealed an enormous amount of variation from one population to the next. Five of the populations had actually evolved to be worse—sometimes much worse—at using maltose than their ancestor. For these populations, adaptation to using glucose had come at the expense of being less able to utilize maltose. And remember, during the course of the LTEE experiment, the populations never actually were exposed to maltose—the decreased ability to use maltose was just an incidental consequence of the changes that had gone on during adaptation for increased glucose intake. Conversely, the other seven populations had gotten better at their ability to use maltose.

What this means is that by generation 2,000, the twelve LTEE populations were quite variable genetically. Even though their growth rate on glucose was approximately the same from one population to the next, this uniformity cloaked an underlying heterogeneity in the genetic differences that had evolved among the populations.

Since Travisano published this study, a number of conceptually similar projects have been conducted, with much the same result. Even though replicated populations seem to adapt in similar ways when exposed to the same selective conditions, placing them in novel environments reveals cryptic genetic variation leading to heterogeneous responses to the new conditions. In other words, appearances can be deceiving—evolution in identical environments from the same starting point is not as deterministic as it might seem!

It's not much of a step to get to the next question. Suppose populations evolve for many generations in one environment. Will the differences that accrue during that time affect not only how they initially fare in a new environment, but also how they subsequently adapt to the new conditions? Few studies have broached that question and Travisano's is still the benchmark.

After discovering the heterogeneity in the initial response of the LTEE Twelve when placed in a maltose environment, Travisano allowed the populations to adapt to this new resource. This experiment was conducted in exactly the same way as the LTEE, except that the nutrient provided in the medium was maltose rather than glucose. And just like the LTEE, the populations adapted over time—one thousand generations later all populations were better adapted to using maltose than the original ancestor had been. Moreover, the degree of adaptation was related to each population's initial fitness—those that started out doing poorly in maltose experienced a much greater increase in adaptation than the populations that initially fared well. Indeed, the effect was so great that by the end of the experiment, the populations were all nearly equally fit on maltose—the initial differences in fitness were greatly diminished.

Still, some differences existed—the populations that were best at using maltose at the start of the experiment still grew about ten percent faster than the populations that initially were the worst maltose eaters. The results for cell size were similar. There was some semblance of a trend—the two populations that initially had the smallest cells experienced the biggest increase and the population with the largest cells had the biggest decrease. But there was also a lot of inconsistency, with some populations that started with similar cell sizes evolving in different ways.

In other words, the initial variation among populations in degree of adaptation to using maltose—accidentally evolved while the LTEE populations were evolving in glucose—had a long-lasting effect. One thousand generations of adaptation to maltose had failed to erase the signal of genetic differentiation.

In a Gouldian sense, this result is profound. Even when populations evolve in parallel, the hidden differences that are accruing may steer them in different directions should they be exposed to novel conditions.

WONDERFUL LIFE has had a tremendous scientific impact. Despite being written for the general public, the book has been cited in nearly four thousand scientific papers, an enormous number (scientists are usually pleased when their work accumulates fifty or a hundred citations). "Replaying the tape of life" has become part of the lexicon, stated without explanation because everyone knows what the phrase means.

Yet, although John Beatty is too courteous to say it directly,* Gould really messed things up with his metaphor. Gould's instructions to "press the rewind button . . . go back to any time and place in the past. . . . Then let the tape run again" is very clear. But it's not at all what Gould meant. Or, at least, Gould meant a lot more than that.

Two mostly separate research programs have developed to test Gould's idea. Taking him at his word, the LTEE and similar studies have rewound the tape—either literally by resuscitating ancestral populations or by conducting the replays spatially, one flask next to another.

An alternative approach has done what Gould said to do, even if it didn't square with his catchy metaphor. These studies subjected populations to somewhat different conditions to see how resilient evolution is to perturbation. Will evolution always get to the same endpoint regardless or are the results contingent on conditions at the outset and what happens along the way?

It's not really surprising that these two approaches seem to give, on average, different results. If populations start exactly the same and experience the same environments, they usually evolve in more or less the same way. There is randomness in which mutations occur, and that randomness will cause populations to diverge, occasionally a lot, but usually just a little, as long as they remain in the environment to which they've been adapting.

By contrast, if they start differently or experience different events through time, populations are more likely to diverge. Surprisingly few studies have examined this scenario—the one that Gould really em-

---

* I can speak with firsthand authority on this subject, because I was myself a beneficiary of Beatty's too-nice-to-criticize manner. In his paper on Gould's views on contingency, he used my work on lizard evolution in the Caribbean as an example (the other example, incidentally, was the Travisano-Lenski paper on maltose). In his paper, Beatty notes that I wrote something that makes no logical sense. Instead of belaboring the point, however, he writes that I subsequently explained to him what I meant and so it really did make sense after all, despite what was plainly written in my paper.

phasized with his phrasing—but those studies demonstrate that sometimes the evolutionary outcome will be quite different.

Beatty concluded his analysis by suggesting that these two approaches are complementary. The first approach investigates whether populations starting the same will diverge, whereas the second approach asks whether populations starting differently, or experiencing different events, will converge on the same evolutionary outcome.

Alternatively, we could view the first set of experiments as a subset of the second. Even populations starting identically will eventually diverge due to differences in which mutations occur. These differences are as much a result of history as changes driven by external events. Once such genetic differences arise, will they cause populations to diverge further, or will the populations nonetheless continue to adapt in the same general way?

The bigger issue, though, concerns which questions we wish to address. To the philosopher, it's of interest to ask whether populations that start the same and have identical experiences will evolve in identical ways. But to naturalists, astrobiologists, and—I contend—Stephen Jay Gould, the question is different. In nature, populations never start out exactly the same and they never experience the same sequence of historical events. When such populations occur in similar environments, is natural selection all-powerful, as Conway Morris and others suggest, enough to override the different genetic constitutions and different histories? Or is selection constrained to work within the confines of the historical record, limited in possibility by what has gone before and thus likely to yield a different outcome every time, as Gould contended?

Laboratory evolution experiments have done a brilliant job of clarifying these questions and showing us the range of different evolutionary possibilities. But their major advantage is also their major drawback: they're limited to the artificial confines of the laboratory. By design, laboratory experiments are beautifully controlled, one flask identical to the next. Extraneous causes—noise in the system—are excluded as much as possible, allowing the investigation to focus on the factors under study. This is all well and good, essential for a well-run experiment.

But as we've already seen, nature is noisy and uncontrolled. The idea of completely identical environments is laughable—the wind blows, insects fly in, a bird overhead poops out a seed that germinates. Messy for sure, outrageous to the lab scientists used to controlling everything. But that's nature.

These are just the sorts of jots and tittles that Gould was talking about: one replay not quite the same as the other due to some event or condition—a seed, a storm, an asteroid. If only we could take the power of microbial evolution experiments and wed them with the happenstance of the natural world—then we could really test the role of contingencies of time and place. It turns out that we can do just that, and at the same time learn about how the evolution of microbes affects human welfare.

chapter twelve

The Human Environment

P*seudomonas aeruginosa* is a wily bacterium, broadly distributed in the environment and highly adaptable: it can thrive in oil spills and while floating around in a space shuttle. It infects plants, nematodes, fruit flies, fish, and many mammals. In humans, it's responsible for infections in burns, wounds, the urinary tract, and the eyes.

What *P. aeruginosa* really likes is moist places, and that makes human lungs an attractive landing spot. For most people, this is not a problem—we just cough them up and spit them out. But it's a different story for those afflicted with cystic fibrosis. CF sufferers have abnormally thick mucus, which makes clearing the lungs difficult. *P. aeruginosa* and other bacteria take advantage of the clumps of mucus, forming mats (called "biofilms") that insinuate themselves into mucoid nooks and crannies, making them hard to eradicate. The result is infections, pneumonia, lung damage, and often death—*P. aeruginosa* is implicated in the mortality of eighty percent of individuals with cystic fibrosis.

Around the year 2000, doctors realized that the story involved more than *P. aeruginosa* colonizing CF individuals and burrowing into the lungs' passageways. Rather, the microbe's lethality is due in part to post-colonization evolution as it adapts to its new pneumatic habitat in ways that make it hard to remove and increase its injurious effects.

This discovery changed the way people with CF are treated. In times past, individuals with CF had been brought together, often going to specially tailored CF summer camps and hospital wards. Now we realize that this is the worst possible approach; such congregations helped transmit highly evolved, virulent strains of *P. aeruginosa* from one individual to another. Now, contact between individuals with CF is prevented as much as possible, especially in hospitals.

Consequently, today most people with CF get *P. aeruginosa* not from others with CF, but from environmental sources. From the microbe's perspective, each CF individual is an opportunity, each *P. aeruginosa* colonization an evolutionarily independent event. And that, of course, leads to a now-familiar query: do *P. aeruginosa* strains, when adapting to similar, but not identical, environments—in this case, human lungs—evolve in similar ways?

In theory, this question could be tested in the lab, just like the many microbial experimental evolution studies discussed in the last three chapters. Indeed, such an experiment has been conducted. Several enterprising Canadian researchers created faux human lungs, fabricating a viscous, gooey substance meant to be similar to the mucus in a CF individual's lung. They then placed *P. aeruginosa* in petri dishes full of the goop and watched them adapt.

Like most microbial evolution experiments, the replicate *P. aeruginosa* populations showed many similarities in the ways they adapted to the novel environment. But all of the populations were drawn from microbes in the same petri dish culture—they all started out genetically similar. By contrast, the *P. aeruginosa* strains in CF patients probably differ substantially from one individual to another.

In most cases of CF infection, we don't know the environmental source of the *P. aeruginosa*. Very possibly, the colonization route is fortuitous—perhaps one individual gets her microbes from a water faucet, another from a bird-watching trip to a swamp. These *P. aeruginosa* strains, potentially adapted to very different environments, are likely to vary genetically. And as we know from previous studies, when an experiment is started from populations with heterogeneous evolutionary and ecological backgrounds, the populations may adapt in distinctive ways. So, it's not obvious that we should expect diverse CF strains to adapt to different people in the same way.

One thing is certain, however: bacteria invading the respiratory tract of someone with CF will find conditions quite different from the outside world. Not only will there be an active immune system fighting the invaders and antibiotics trying to disable them, but the *P. aeruginosa* also must contend with a variety of other competing bacterial species and the sticky mucus coating. Natural selection pressures will be very intense.

Furthermore, the human respiratory system provides many different environments, from the nasal sinuses to the respiratory bronchioles and alveoli. Consequently, many different niches are available—differing in airflow, humidity, oxygen content, surface structure, mucus abundance, and antibiotic concentration. That variation, as well as the differences that exist from one human to the next, may lead different strains of *P. aeruginosa* to adapt in different ways, not only between CF patients, but even within them.

How different strains of *P. aeruginosa* adapt, of course, is not an academic question. Most tests of evolutionary repeatability in the lab are done for curiosity's sake. Whether yeast from an oak tree and a woman's reproductive tract adapt similarly to living in a petri dish full of glucose may be of interest to evolutionary biologists, but whether bacteria adapt in the same way in the lungs of people with cystic fibrosis has real-world consequences—the more repeatable the bacteria's evolution, the easier it may be to develop new drugs and therapeutic treatments.

In a world lacking ethics, researchers would intentionally infect CF individuals with different strains of *P. aeruginosa* and carefully monitor the evolution of the bacteria. In the real world, of course, to even contemplate an experiment like that would be abhorrent. But essentially the same set of events goes on all the time as people with CF are attacked by *P. aeruginosa*.

Just such a natural experiment was studied by researchers at the Copenhagen Cystic Fibrosis Center in the early years of this century. As part of the treatment protocol at the center, individuals with CF come in monthly to produce sputum samples that are examined for the presence of *P. aeruginosa*. Those who test positive are immediately put on a treatment regimen that sometimes is effective at eradicating the bacteria.

Although established for therapeutic purposes, these procedures produced an elegant evolution study. The center's clinicians detected infection by *P. aeruginosa* almost immediately after it occurred and then monitored and resampled periodically for up to ten years. By comparing samples taken from the same patient at different times, the staff could chart the bacteria's evolutionary progression.

The Danish researchers sequenced the entire genome of more than four hundred *P. aeruginosa* samples taken from thirty-four children and young adults. In several cases, the strains in different individuals were very similar, suggesting that the bacteria had been passed from one patient to another, despite the clinic's best efforts to prevent such transmission.*

The vast majority of the bacterial genomes, however, were very dif-

---

* In support of this possibility, visit records indicated that the patients in question had been at the clinic at the same time before one of the patients was infected.

ferent from one another, indicating that the patients had independently acquired their *P. aeruginosa* from different environmental strains. The question, then, was how similar were the evolutionary pathways taken by these different bacteria.

By comparing the DNA of *P. aeruginosa* in an individual at different points in time, the researchers could chronicle the genetic changes that had occurred after the bacteria had colonized that person. In total, they found more than twelve thousand mutations, an average of more than three hundred per colonizing strain.

The problem was making sense of this multitude of data. Which changes represented adaptation to the newly occupied human lung environment and which were random changes with no adaptive significance?* The genome of *P. aeruginosa* contains more than five thousand genes and six million bits of DNA. Although progress has been substantial, we still have very limited understanding of how the bacteria's genome works. Thus, the Danish researchers had little idea of the consequences of almost all of the twelve thousand genetic changes they detected.

Faced with this dilemma, the researchers had a stroke of inspiration. They reasoned that convergent evolution of populations occupying similar environments is strong evidence of adaptive evolution. Moreover, microbes are known to convergently use the same genes to adapt to similar circumstances. In the quest to identify genes involved in *P. aeruginosa* adaptation to living within humans, why not try looking for those that are repeatedly mutated in different CF patients?

The researchers cataloged the mutations, tabulating the number of strains that had mutations in the same gene. In total, mutations occurred in nearly four thousand genes, a third of which had mutations in multiple strains. Of course, two strains might gain mutations in the

---

* Many mutations have no effect on the phenotype and are of no consequence.

same gene just by chance. Statistical analysis set the threshold at five—the occurrence of mutations in the same gene in that many strains was extremely unlikely to occur randomly.*

Fifty-two genes gained mutations in five or more strains—the record was one gene in which twenty strains, more than half of the total, experienced a genetic change. The researchers considered these fifty-two genes as candidates for possible convergent adaptation—"candidate pathoadaptive genes in which mutations optimize pathogen fitness," in their parlance.

One way to test the efficacy of this method was to see whether it detected genes already known to be involved in *P. aeruginosa* adaptation. And, indeed, half of the genes they found were ones that had previously been identified, most notably genes involved in the evolution of antibiotic resistance and the formation of biofilms. Relying on convergence does, indeed, detect genes involved in pathogen adaptation.

The promising result from this study is that it identified a number of genes not previously implicated in CF adaptation. The biochemical function of seven of these genes already was known, so investigation now focuses on how altering these functions by mutation can allow *P. aeruginosa* to adapt to CF patients. In addition, nineteen of the convergent genes were terra incognita, their function completely unknown (not surprising given that we don't know how nearly half of *P. aeruginosa*'s genes work). Clearly, if we don't know what a gene does, we can't have any idea how changes in that gene might lead to adaptation to the human lung environment. Figuring out how these genes function is obviously a high priority.

I'd love to be able to conclude this story with the headline "Conver-

---

* This is a slight simplification because the actual cutoff varied depending on the size of the gene.

gent Evolution Saves Cystic Fibrosis Patients," but we're not there yet. Nonetheless, it's clear that studying convergent evolution is of more than just academic interest—it can help us get a handle on how pathogenic (disease-causing) organisms attack humans and perhaps how we can design therapies to combat them.

At the same time, however, the results of this study bear on the question of evolutionary predictability and contingency. Most of the identified genes experienced mutations in fewer than half of the thirty-four patients. Moreover, by relying on convergence, the analysis was unable to identify adaptive mutations that occurred in only one or a few patients. The overall repeatability by which *P. aeruginosa* adapts to CF patients is relatively low. Whether this non-repeatability is the result of the randomness of mutations, the different genetic constitutions of the different infecting strains, biological differences among the patients, or adaptation to different parts of the lungs remains to be determined.

Another study yielded very similar results. *Burkholderia dolosa* was a microbe new to science when it afflicted CF patients at a Boston hospital in the early 1990s, eventually infecting thirty-nine individuals. Just as with the *P. aeruginosa* study in Denmark, repeated samples from the same patients allowed the researchers to track genetic changes of the microbe within each individual.

As with the *P. aeruginosa* study, there were a large number of mutations. Given the obscurity of the bacterium, figuring out the consequences of most changes was difficult, so a team led by Tami Lieberman at Harvard Medical School looked for genes repeatedly mutated in multiple CF patients. The seventeen genes they identified included eleven genes known to be related to antibiotic resistance and disease development. But the function of three of the multiply mutated genes was completely unknown and another three had never been implicated in lung disease. Without this information, no one would have thought to consider them as important in *Burkholderia* infection. Research is currently under way on several of the mutations to understand how they might be pathological.

Also like the *P. aeruginosa* study, relatively few genes were mutated in even half the patients, and thus the overall predictability was again low. And by focusing on genes convergently mutated in multiple patients, the study was unable to identify adaptive changes that occurred in only one or few patients.

Both of these studies were conducted on cystic fibrosis because CF patients are prone to infections. Routine monitoring provides researchers with samples from the earliest stages of infection, which allows them to examine how the bacteria adapt over time. For most diseases, however, multiple samples from each individual are not available. In many cases, such samples wouldn't even be useful, because many bacteria arrive in their latest victim with their pathogenic adaptations already in place, having evolved elsewhere.

An alternative means of identifying convergent genetic changes is to follow the lead of evolutionary biologists and build a phylogeny to examine trait evolution. By comparing virulent strains with inoffensive relatives, medical microbiologists look for similar changes that have evolved multiple times only in the pathogenic strains.

Most often, researchers have used this approach to examine the genetic basis of drug resistance. For example, *Mycobacterium tuberculosis*, the bacterium responsible for tuberculosis, has evolved antibiotic resistance many times. An international team of researchers sequenced the genome of 123 strains of *M. tuberculosis*, forty-seven of which displayed resistance to antibiotic drugs used to treat TB. As expected, the resulting phylogeny confirmed that the antibiotic-resistant strains were not all closely related to each other; rather, resistance to the drugs had evolved convergently many times.

Across the entire sample, the researchers identified nearly twenty-five thousand DNA positions that had experienced a mutation in at least one of the strains. The researchers then focused on the mutations that evolved multiple times and only or primarily in strains that were resistant. The extreme case was one mutation that had independently evolved in eight resistant and no non-resistant strains.

The study was wildly successful. Eleven regions of the *M. tuberculosis* genome—either genes or DNA occurring between genes—had previously been identified as having mutations that conferred antibiotic resistance. All eleven were identified in the study. But in addition, another thirty-nine regions were found that had not previously been suspected of involvement with TB. Eleven were in genes whose function is already known. Several of these genes are involved in determining the permeability of the bacterium's cell wall, suggesting that such changes may in some way be involved in antibiotic resistance, perhaps by making it harder for the antibiotic to enter the bacterium's cell. The remaining twenty-eight changes were in genes of unknown function. Research is currently under way to better understand how these changes lead to antibiotic resistance and, ultimately, how such evolution may be prevented or countered.

THE EXTENT OF CONVERGENCE in these studies was far from universal. Even the most extreme cases of convergence involved barely more than half of the strains; most convergently mutated genes only occurred in a distinct minority of the strains. In fact, most genes were only mutated in a single strain. In the convergence-versus-contingency dispute, these data seem to fall solidly in Gould's favor.

For biomedical practitioners, this debate misses the point: some predictability is immensely better than none at all. Even if not all microbes adapt using changes to the same gene, the fact that some do evolve in the same way is important information. By learning the mechanism of adaptation, we can devise medicinal countermeasures to deploy in those cases when the focal gene is involved. By taking samples from the patient, we can quickly sequence the microbial genome and find out whether the infecting strain has the genetic change in question. If it does, unleash the therapeutic hounds. If not, look for other possible causes. Roy Kishony, whose work has been centrally important on this topic, expressed this view more technically, writing (in collaboration with graduate student Adam Palmer) that "even a modest predictive power might improve therapeutic outcomes by informing the selection of drugs, the preference between monotherapy or combination therapy and the temporal dosing regimen to select genotype-based treatments that are most resilient to evolution of resistance."

This is one aspect of the much-ballyhooed "personalized medicine" in which physicians can identify the particular cause of an individual's illness and then treat accordingly. And the fact that some microbial pathogens evolve convergently makes this approach much more feasible.

COMPARATIVE STUDIES of pathogen strains aren't the only approach being taken to understand microbial evolution and its effect on human health. Evolution experiments, so valuable for enlightening our understanding of microbial evolution, are also being used to search for predictable ways that microbes adapt to attack us and foil our countermeasures.

Most of these studies examine the evolution of antibiotic resistance using the general approach pioneered by Lenski, Rainey, Travisano, and others: confront microbes with various insults and watch them adapt. At the most basic level, these studies look for repeated evolutionary patterns. If microbes evolve resistance in the same way time and time again, then researchers can focus their efforts on thwarting that specific evolutionary response.

A particularly clear example of the value of such work came from Kishony's lab at Harvard Medical School, the same place where the *Burkholderia dolosa* research was conducted. In this experiment, our old friend *E. coli* was placed in specially designed growth chambers and exposed to one of three antibiotics—chloramphenicol, doxycycline, or trimethoprim—and its evolutionary response followed for twenty days (about 350 *E. coli* generations). Each treatment was replicated five times.

The goal of the study was to observe the evolution of resistance to the antibiotics. Initially, the bacteria—all derived from the same progenitor—were not resistant and grew very poorly in the presence of the antibiotics. But very quickly, resistance began to evolve and growth rates increased.

The microbe populations showed very similar patterns of adaptation to the drugs. For all three, the five replicate populations steadily increased in resistance—as much as a 1,600-fold increase in the populations exposed to chloramphenicol. At the end of the experiment, the researchers sequenced the genomes of cells from each of the fifteen populations and compared them to the genome of the ancestral population.

In line with most previous experimental studies of microbial adaptation initiated with identical strains, the five populations exposed to trimethoprim evolved in very similar ways. Trimethoprim works by disabling the *dihydrofolate reductase* (*DHFR*) gene in *E. coli*. Thus, it's perhaps not a surprise that the counterstrategy of *E. coli* was to modify *DHFR*, making it harder for the drug to recognize the gene and boosting the production of the enzyme the gene produces. Almost all the changes in the five populations were in *DHFR*. In total, seven different mutations were detected in this gene: one of these mutations occurred in all five populations, another in four populations, and all but one mutation occurred in at least two populations. Other than mutations in the *DHFR* gene, only three additional mutations occurred, each in a different gene and each occurring only in one population.

Given the high level of repeated evolution of particular mutations, the researchers sequenced samples of *DHFR* from each population from every day of the experiment; they discovered that there was a consistent order in which mutations occurred, with the same or similar-acting mutations invariably preceding others. In other words, the evolution of trimethoprim resistance in *E. coli* is highly repeatable.

The results were very different for the populations exposed to the other two antibiotics. Even though the degree to which resistance evolved by the end of the experiment was similar among the five replicates for each drug, examination of the genetic changes revealed mostly different mutations had evolved in each population.

Why *E. coli* evolves repeatedly in similar ways to one drug, but unpredictably in response to two others, is unclear. Regardless, the results indicate that devising general solutions to the problem of antibiotic resistance will be easier for trimethoprim than for the other two drugs.

I'VE PREVIOUSLY MENTIONED that some scientists don't like the messiness of research beyond the confines of lab walls. Too much environmental noise, too many confounding, uncontrolled variables. This concern is particularly apt with regard to convergent evolution—if environments aren't the same, then a lack of convergence may simply be the result of different selective pressures. A recent study of stream stickleback fish found just this. At first, University of Texas researchers were puzzled by the lack of convergence among populations that had independently colonized different waterways. But when they looked more carefully, the explanation became clear: variation in water quality and vegetation among streams could account for phenotypic differences and non-convergence among the fish populations. Of course, the same explanation—subtle environmental differences—could account for lack of convergence in strains of *M. tuberculosis* or *P. aeruginosa* inhabiting different individuals or, for that matter, in any case of non-convergence.

Some scientists are even suspicious of the lack of convergent responses in controlled laboratory studies. Perhaps even the slightest difference from one test tube to another—a fraction of a degree in temperature or a minutely greater amount of sunlight from a nearby window—could be enough to lead to different selection pressures and hence, non-convergent adaptation.

But these laboratory-based skeptics pose a deeper critique of the convergent evolution approach to the study of evolutionary predictability, relating to an important question that I have ducked to this point. So far, I have used the terms "repeatability" and "predictability" pretty much interchangeably. But are they the same thing? And more to the point, just because convergent evolution is the phenomenon of repeated evolution, does that make it a good way to study evolutionary predictability?

Some don't think so. A pair of European scientists, for example, wrote that repeatability "is a weak form of predictability, as the deterministic nature of the process can be ascertained only in retrospect." In other words, true prediction occurs a priori, based on a detailed understanding of the system under study, rather than simply seeing what happens repeatedly and predicting that it will do so again.

These scientists wouldn't be satisfied with observing that small size repeatedly evolves in elephants confined to islands; rather, they'd want to be able to predict evolutionary diminution from understanding how insular environments affect the evolution of body size. Even in laboratory evolution experiments, the observation that cell size always increases or that the same gene incorporates mutations when populations are exposed to the same conditions is not sufficient. They want to be able to specify the expected outcome before the experiment is run.

At the macroscopic level, scientists make such predictions all the time. This approach from first principles is what Dale Russell did in hypothesizing the dinosauroid. Based on his understanding of anatomy, he was able to predict how selection for larger brains in a theropod dinosaur would lead to other anatomical changes, ultimately resulting in an organism very human-like in appearance.

In a much more sophisticated way, researchers in the fields of physiology and biomechanics have long studied the relationship between anatomical design and organismal function. What is the best wing shape for a bird that needs to maneuver sharply? Short and stubby, just like a fighter jet. What are the best body proportions for living in cold places? Stocky, with short appendages to minimize surface area through which heat can be lost.

These are predictions made independently of what has actually evolved. Subsequently, they can be cross-checked with nature. In some cases—such as the two I've listed—the predictions are borne out: natural selection does seem to favor optimal solutions. In other cases, theory and nature don't jibe—either the theory was off or some constraint has prevented natural selection from sculpting the optimal solution. What those constraints might be is an interesting topic in itself: perhaps suitable mutations don't occur or tradeoffs get in the way (it's not possible to simultaneously optimize everything). Or perhaps the solution is simply impossible—no organisms use nuclear fission as an energy source, for example, and biological wheel-like structures are exceedingly rare.

It's more difficult to make predictions from first principles when dealing with microbial organisms because the biochemical and molecular workings of these cells are still not well understood. This difficulty is compounded when working at the genetic level—as most microbial workers are these days—because the function of most genes remains a mystery. Consider all the genes identified in the TB and CF surveys whose function is completely unknown—it would be pretty hard to make a priori predictions about how they might be involved in adaptive evolution by microbial pathogens.

Of course, some exceptions exist. One is a gene responsible for antibiotic resistance in *E. coli*. The beta-lactamase gene produces an enzyme, beta-lactamase, that evolves to attack antibiotics such as penicillin, ampicillin, cefotaxime, and many other antibiotics, rendering them ineffective. For this reason, the beta-lactamase gene and the enzyme it produces have been intensively studied and are much better understood than most microbial genes and their products.

A recent study investigated the diversity of mutations that occur in the gene. Through molecular trickery, the researchers caused *E. coli* cells to produce ten thousand different mutations. They judiciously chose a thousand and measured their effect on antibiotic resistance (quantified by how much antibiotic was required to kill the cell). Some mutations made no difference, a few mutations were catastrophic, and most had an intermediate, mildly detrimental effect.

Because beta-lactamase is so well studied, the researchers were able to determine how each mutation affected the functioning of the enzyme in terms of the manner in which the molecule changed its shape, its activity level, and how stable it was. They could then correlate these changes with the effect on antibiotic resistance and found that there was a strong relationship: bigger changes were correlated with greater alterations in resistance. In other words, the researchers were able to start with a mutation, figure out how the mutation modified the enzyme it produced, and from those changes accurately estimate how antibiotic resistance would be affected. It is just this sort of approach that could allow researchers to predict how a microbe like *E. coli* would evolve when faced with new environmental circumstances.

But examples like this are much more the exception than the rule. In most cases, we don't know which genes are responsible for adaptation. Even if we know which genes are involved, we often have little understanding of how the genes work, much less the effect that particular mutations have. Perhaps someday we routinely will be able to predict which mutations will evolve adaptively, but that day is a long way off.

In the absence of such comprehensive information, researchers sometimes use more incomplete data to make predictions. For example, researchers at Harvard noted that one strain of *E. coli* is capable of withstanding doses of the antibiotic cefotaxime one hundred thousand times greater than needed to knock out non-resistant strains. Genetic analysis demonstrated that this high level of resistance was the result of the acquisition of five mutations in the beta-lactamase gene.

The researchers focused on these five mutations and investigated whether, starting with the non-resistant strain lacking any of the mutations, natural selection inevitably would lead to the five-fold mutated strain. Rather than conducting evolution experiments, though, they created strains of *E. coli* with all possible combinations of the five mutations. For each strain, they measured resistance to cefotaxime and asked, "For each strain, is there a mutation that if added would increase resistance?" For example, for a strain with two of the mutations, would addition of any of the other three mutations increase resistance? For all strains, the answer was yes. All strains with one mutation would eventually add a second, all with two would add a third, and so on. Regardless of the order in which mutations appeared, the five-mutation strain was the inevitable outcome. The authors concluded that the "tape of life may be largely reproducible and even predictable."

The study—elegant and comprehensive—received a lot of attention as an example of evolutionary determinism at the genetic level. But there was one problem: the study had been restricted to only those mutations found in the ultra-resistant strain. What about other mutations? Could they throw a spanner into the works?

To find out, a group of Dutch researchers conducted an evolution experiment, which meant that the universe of mutations was defined by whatever occurred during the experiment, rather than being limited to the focal five from the previous study. In the free market of mutations, was the super-strain sure to evolve? The Dutch team designed their experiment in the now-familiar way, exposing twelve initially similar populations to the drug for several generations and quantifying the extent of adaptive evolution.

Resistance to cefotaxime increased over the course of the experiment, but its extent varied, with seven populations becoming substantially more resistant than the other five. The scientists sequenced the genomes of each population and discovered that the same three mutations—three of the super five—had evolved in essentially the same order in the seven highly resistant populations.* By contrast, at least one of these three mutations had failed to evolve in the other five populations.

The Harvard super-strain study had shown that one particular mutation—G238S in microbiology-speak—is the single most effective mutation in conferring resistance to cefotaxime. In the Dutch study, all seven of the high fliers had immediately evolved G238S, as had three of the laggard populations. The Dutch researchers scrutinized the two populations that had failed to evolve G238S and identified the first mutations they had acquired, R164S in one of the populations and

---

* As for the other two of the super-five mutations, the evolution of one was precluded due to technical reasons inherent in the experiment's design, but the fifth mutation's failure to evolve was a mystery.

A237T in the other;* neither of these mutations had evolved in any of the other ten populations. Moreover, because neither of these mutations was among the five found in the super-resistant strain, they hadn't been included in the Harvard team's study.

The Dutch team then started the experiment again, but this time they began with *E. coli* that possessed one of these two mutations, five populations with R164S and five with A237T. Again, cefotaxime resistance increased through time, but all ten populations ended up with resistance levels substantially below that of the seven most resistant populations in the first experiment. Notably, none of these populations evolved G238S, but they did incorporate many other mutations not seen in the populations with G238S.

Why G238S is incompatible with R164S and A237T is not entirely clear, but it appears that the mutations cause an enzyme to fold in different ways; once the first mutation changes the folding pattern, the second mutation would cause disruptive changes in the new configuration, and thus the mutations, although individually favorable, cannot occur in combination. It's like origami: once you start down the path to making an elephant, you can't change mid-course and make a goldfish.

The Dutch study is a perfect example of a historical contingency, a chance event that radically shapes the subsequent evolutionary outcome. Populations in which, by chance, G238S crops up first can go in one direction and often evolve high levels of resistance. But those in which the other mutations happen to occur first are precluded from this path—once they are established, G238S is no longer beneficial and instead adaptive evolution takes a different road, one that leads to a

---

* These names describe amino acid position changes. The first letter is the ancestral amino acid, the numbers referring to the position in the gene, and the second letter is the new amino acid caused by the mutation.

more inferior, less resistant destination. The Harvard team's experiment had not been set up to look for other genes, and thus they didn't discover the extent to which adaptation to cefotaxime was unpredictable.

The contrast between the approaches and results of the Harvard and Dutch studies exemplify why it is so hard to make a priori predictions of evolution at the genetic level. The genome is just too big and complicated to isolate all the relevant mutations and forecast which ones will affect each other and how. Just because a certain set of mutations leads to a highly adaptive outcome doesn't mean that those mutations necessarily will evolve. Often, there are many different ways to produce the same genotype—recall the sixteen different genetic pathways to produce a wrinkly spreader in *P. fluorescens*—and just as many different solutions to the same environmental problem. Figuring out in advance which is most likely to occur and which is unlikely is in almost all cases beyond our capabilities.

A lot of very smart people are working on this problem—both at the molecular and theoretical levels—so perhaps, like the weather, our ability to predict tomorrow's evolutionary forecast will improve; at the moment, however, our capabilities are limited. And that, in turn, means that the best way to predict what will evolve will be to look at what's happened in the past, either through evolutionary time or as the result of evolution experiments.

THE RESULTS FROM STUDIES of microbial adaptation indicate that some degree of predictability exists and that this repeatability can be the basis for the development of countermeasures. Of course, microbes aren't the only contemporary organisms evolving to our detriment. Weeds invading our lawns and agricultural fields, insects and rodents eating our crops, mosquitoes transmitting diseases—all have one thing in common: they've evolutionarily outmaneuvered our attempts to control them.* And just as with the microbes, their cost is measured in billions of dollars and tens of thousands of lives.

The evolution of resistance to pesticides (construed broadly to include insecticides and herbicides[†]) shares many parallels with the evolution of antibiotic resistance. Like many microbes, pests have evolved a wide variety of ways to defeat our chemical arsenal, including changes in behavior that minimize contact with the pesticide; alterations in the exterior skin to keep the pesticide out; development of means to convert the pesticide to something else, sequester it in an unimportant part of the body, or quickly excrete it; or modifications in the molecular structure targeted by the pesticide. Because of these myriad possibilities, populations of the same species often adapt in different ways when exposed to a particular pesticide.

On the other hand, many pesticides are commercially successful because they attack biochemical pathways shared by many pests. As a result, many species have evolved similar—oftentimes identical—means of foiling these attacks. For example, a number of mosquito species have evolved the same change in DNA to adapt to the insecticide dieldrin. Similarly, more than thirty different insect species—including flies, fleas, cockroaches, moths, thrips, aphids, beetles, and kissing bugs—have acquired the same DNA change to become resistant to pyrethroid pesticides.

As with microbes, when pests evolve convergent mechanisms of pesticide resistance, our ability to fight back is enhanced. Pesticides derived from the bacterium *Bacillus thuringiensis* are a good example. For

---

* For example, nearly six hundred species of arthropods have evolved resistance to one pesticide or another.

† And to those more technically minded, also fungicides, larvicides, rodenticides, molluscicides, acaricides, and any other vermin-icide you can think of.

reasons unknown, this soil microbe produces proteins that are lethal to insects. Scientists have identified these proteins and used them as insecticides. Initially, these Bt insecticides, as they are called, were sprayed on crops, but since the late 1990s, several crop species have been genetically engineered to produce the proteins themselves. The amount of farmland now planted with Bt crops is extraordinary, two hundred million acres worldwide in 2013, including two-thirds of all corn in the United States and more than three-quarters of the cotton in the major producing countries.

Resistance to Bt toxins has evolved readily in laboratory experiments and, to a lesser extent, in the field. Bt toxins work by binding to proteins in the gut of insects. Resistance evolves predominantly by mutations that interfere with production of these binding proteins. For example, resistance to one type of Bt toxin has evolved in many populations of three caterpillar species as a result of mutations in a gene that produces a toxin-binding protein called cadherin. Similarly, seven caterpillar species have convergently evolved resistance through mutations that disrupt a gut protein that transports molecules across membranes.

Recognition that mutations in a few genes evolve repeatedly has important implications for combating the evolution of resistance in several ways. First, pest populations can be screened regularly to look for the appearance of specific resistance mutations. These screens involve methods to detect mutations identified in the field or in lab-selected populations. When such alleles are detected early, management actions can be taken to prevent the mutation from becoming widespread.

More generally, the discovery that populations repeatedly evolve resistance in the same way can spur efforts to genetically modify the Bt gene in crops to circumvent this mechanism. For example, once researchers realized that insects were evolving resistance by preventing binding to cadherin, they modified the Bt toxin so that it bound to other proteins, bypassing cadherin entirely.

This is not to say that convergence is a magic bullet. Even in cases like Bt toxin, screens are only effective at finding the previously detected convergent mutations. Other mutations at the same gene may not be spotted, much less mutations involving other genes and resistance mechanisms (in fact, non-convergently evolved mutations in other genes have been reported, as well as many other mechanisms of Bt resistance). And for many other pesticides, the knowledge that resistance has evolved convergently may not lead to new approaches.

OUR EFFECTS ON THE ENVIRONMENT go well beyond the use of antibiotics and pesticides—we're changing the world in countless ways. Sometimes, the challenges we're posing are too great and species decline and go extinct. But in many other cases, natural selection is kicking in and species are adapting to their new circumstances.

Convergent evolution due to human-caused change was first identified as a response to our pollution of the environment. Plant adaptation to soil infused with toxic levels of heavy metals and moths evolving dark coloration in polluted areas were two early examples and more continue to be documented. One particularly well-understood case refers to a small fish found in marine estuaries along the Atlantic coast of North America. The Atlantic killifish, a distant relative of the species studied in Trinidad by Endler and Reznick, is able to thrive in highly polluted areas that few other species can tolerate. A team led by University of California scientists examined four pollution-tolerant populations spread across the East Coast and determined they had independently modified the same physiological pathway, rendering them insensitive to even extremely high levels of many pollutants, including dioxin. Genomic analysis indicated that mutations in the same set of genes were instrumental in this adaptation in all four populations.

Humans also impose strong selective pressure when we remove

animals from populations for commerce or sport. In many cases, hunters target individuals with particular traits. The result is strong selection against individuals with that trait, and in many cases populations convergently evolve similar responses. Trophy hunters, for example, prefer the largest, most magnificent specimens. Consequently, it is no surprise that small ornaments and weapons have evolved in many species, including smaller horns in bighorn sheep and sable antelopes, reduced antlers in deer, and diminished tusks in elephants. Indeed, some elephant populations now contain many completely tuskless individuals.

The same phenomenon is seen in our fisheries. Many means of catching fish are size-selective: most nets, for example, ensnare larger fish, but let the smaller ones slip through. The result is a selective advantage to the little guys. Consequently, the maximum size of many different fish species is only a fraction of what it used to be. The largest Atlantic cod in Canada's Gulf of Saint Lawrence, for example, decreased from seventy pounds in the early 1970s to twelve pounds today; cod off the coast of Massachusetts are about the same small size today, dwarfed by the two-hundred-plus-pound fish caught there in the late nineteenth century.* This is a major economic problem because the number of fish in a population doesn't increase to compensate for the smaller size of the individuals. The result is that fishing yields—the amount of tonnage caught by the fleet—invariably decline.

A critical question is whether populations will convergently return to their ancestral condition if we restore the environment to its original state. In some cases they do; for example, the peppered moths in many

---

* Scientists debate the extent to which these reductions—both in body size and in particular structures like tusks and antlers—represent evolutionary change. The act of removing the largest individuals with the largest structures guarantees that the survivors will be less well-endowed even if no genetic change occurs; in addition, phenotypic plasticity may be partly responsible as well. Nonetheless, at least in some of these cases, a genetic basis for these reductions has been established.

places re-evolved their speckled countenance once air pollution was eliminated. In other cases, however, responses are less consistent. Once size-selective fishing and hunting are curtailed, for example, fish commonly do not re-evolve larger size and bighorn sheep don't re-evolve larger horns. There are a number of possible explanations for this evolutionary asymmetry. Possibly, selection for large size in the absence of harvesting is much weaker than selection for small size was during harvesting. Alternatively, harvesting may have driven the ecosystem into a new equilibrium in which large size is no longer favored. For example, other species may have expanded their populations to take over the resources formerly used by the harvested species, in which case selection pressures might be permanently changed even after harvesting ceases.

As with pesticides and antibiotics, convergent evolution is only part of the story in understanding how species will respond to a changing environment and what we might do to improve the situation. Nonetheless, when it does occur, convergence clearly frames the problem and invites the development of general counterstrategies. Indeed, to prevent reduction in fish size, scientists have devised a number of approaches, including development of non-size-selective nets, throwing back the very largest fish to maintain their genes in the population, or maintaining fishing-free zones where large fish can thrive and export their genes for large size to fished areas.

No doubt, as scientists increasingly study how species are responding to a globally changing world, more cases of convergent evolutionary response will be detected. The elephant in the room, of course, is global warming. To date, few studies have convincingly demonstrated evolutionary adaptation due to climate change, but this is rapidly changing. I am unaware of examples of convergence in natural populations, but one seven-year-long experimental study on worms detected replicated genetic changes associated with warmer soils. I predict that this is just the tip of the melting iceberg and that soon we will detect many physiological, behavioral, and anatomical changes convergently evolved in vulnerable species.

In contrast with size reduction in overharvested fisheries, the challenge in this case will be to use the information gleaned from convergence not to prevent evolution, but to enhance its effectiveness. Predicting in advance what form such interventions might take is hard, but they might include introducing particularly effective genes into needy populations lacking them and altering the environment in ways to enhance frequently evolved behavioral and physiological adaptations.

More generally, we are on the threshold of a new era in which we have unprecedented ability to direct the evolutionary process. The development of new molecular technologies—most recently and importantly, CRISPR*—has raised the specter of genetically engineering wild populations, of basically being able to direct genetic evolution in the wild. Already, plans are afoot to genetically modify mosquitoes so they can't transmit certain diseases like malaria to humans. This is a brave new world, raising many objections, both practical and ethical. These concerns are well founded, but there is also a potential upside. Not only might we engineer species for our own benefit, but we might be able to help species help themselves, by introducing genes that will allow them to adapt to a changing world.

And how might we know what genes to introduce to a species facing a particular environmental challenge? Convergent evolution, of course! By looking to solutions that have worked repeatedly for other species, we may be able to identify the best candidates for genetic rescue of imperiled species. Whether such a future comes to be remains to be seen, but if it does, convergent evolution is likely to play an important role.

---

* Clustered Regularly Interspaced Short Palindromic Repeats

Fate, Chance, and the Inevitability of Humans

Ten feet tall with long tails and pointy ears, the Na'vi live on a moon near Alpha Centauri. *Avatar*'s blue-skinned humanoids share their world with a lush and biologically rich ecosystem strikingly similar to life here on Earth. Sure, the animals sometimes have an extra set of legs or a face like a hammerhead shark, but for the most part, they resemble creatures with which we are familiar: leopards, horses, monkeys, pterodactyls, titanotheres,* birds, and antelopes. The luxuriant vegetation appears to be straight out of the Amazon rainforest, so similar to earthling plants that a botanist produced a classification complete with scientific names.

The attention to detail and beautiful production are what set *Avatar* apart from other films of this genre and the reason it won Oscars for Best Art Direction, Best Cinematography, and Best Visual Effects. But in terms of biology, *Avatar* is similar to most other movies set on other worlds. From *Star Wars* to *Guardians of the Galaxy* and beyond, most

---

* Enormous, ancient rhino relatives

interplanetary science fiction movies populate their worlds with life-forms quite similar in general appearance and biology to what has evolved here on Planet Earth. Even some of the more bizarre types, such as the fearsome predators of *Dune* and *Alien*, exhibit biology traceable to terran species.

Indeed, for the most part, the only really different cinematic life-forms are based on biology of a very different sort than here on Earth. Instead of carbon as the basic elemental building block, these stories posit life based on silicon or even pure energy, producing species composed of crystals, interstellar protoplasm, or wavelengths of energy.

Astrobiologists think that extraterrestrial life, if it exists, is most likely to be based on carbon and thus to be similar in its chemistry to life on Earth. So let's restrict discussion to carbon-based life-forms potentially arising on the many Earth-like planets that we now know exist elsewhere in the Milky Way galaxy. Is it probable that the ecosystems on these planets will be populated by species similar to those on Earth? Should we expect, as many movies would suggest and Simon Conway Morris actually said, that "what we see here [on Earth] is at least broadly, and I suspect much more precisely, what we will find on any comparable Earth-like planet"?

Conway Morris and others make this claim based on two arguments. First, the pervasiveness of convergence here suggests that natural selection tends to produce the same solution to common problems posed by the environment. Second, the laws of physics are universal, at least in our universe, and they dictate certain optimal ways of adapting to the environment that are not specific to our world. Conway Morris adds a third, more speculative, argument: some biological molecules that have evolved here on Earth, such as DNA, chlorophyll (used by plants for photosynthesis), opsins (the molecules used to detect light in visual systems), and hemoglobin (used for transporting oxygen in blood) may represent the best, or among the best, possible molecules in a system using carbon-based materials. So, it's not implausible, Conway Morris and others posit, that similar fundamental building blocks will arise on other planets.

I would agree that some examples of convergence between extraterrestrial life-forms and species here on Earth are likely. Just like here, organisms elsewhere will need to gain energy, either by producing it themselves or ingesting external resources. They'll have to have sensory capabilities to detect external stimuli. Some will need to move.

Earthly species are pretty good at these tasks, and so it wouldn't be surprising if there were parallels on other planets. This is particularly true because gravity, thermodynamics, fluid mechanics, and other physical phenomena apply everywhere. Organisms that need to move rapidly through a dense medium will maximize their performance by evolving a streamlined body shape. Powered flight requires some means of generating lift—wings do a good job of that. Focusing light is most effectively done with a camera-like structure like the eyes that have evolved many times in the animal world.

As a result, at least some parallels between terrestrial and extraterrestrial life probably will exist, particularly on those planets most similar to Earth. Such convergence might be accentuated if the same molecular building blocks evolve in other life-forms, although the extent to which using the same molecules necessitates phenotypic convergence is not clear.

Despite the occurrence of some cases of convergence, I would expect extraterrestrial life to be for the most part very different from what we see here on Earth. Evolution experiments and a consideration of convergent evolution teach us that distantly related species, when exposed to the same conditions, often evolve in different ways.

And species on other planets would certainly be distantly related. Not only would the life-forms be evolving from different starting points, but even if life were carbon-based and the genetic code were based on something like DNA, the rules of inheritance and evolution might be very different. Maybe characteristics acquired during an individual's life could be transmitted to offspring. The recombining of parental genes that occurs as a result of sexual reproduction might not occur, or it might involve three, ten, or one hundred individuals per mating event instead of our customary two. Natural selection as we know it—based on competition among individuals of the same species for limited resources—might not operate. Perhaps cooperation, both within and between species, might be the driving force in evolution. Discrete species might not even exist. Given such possible differences in the organization of life, it seems unlikely that evolution would follow the same route on different planets.

And the planets themselves would be very different. If Rich Lenski were omnipotent,* he'd simply create a dozen Planet Earths, place them in identical solar systems, wait a few billion years, and then come back and see how life—assuming it had arisen—was similar or different on his identical globes. But until Lenski or someone else figures out how to conduct that experiment, we're stuck with comparing how life evolved on different planets (if, indeed, it has).

We used to think that Earth was unique, that nothing like it existed anywhere else. We now know that we were very wrong. It seems like every week there's another announcement of the discovery of additional habitable exoplanets. Extrapolations from these finds put the number of such planets just in our corner of the universe—the Milky Way galaxy—in the billions.

Keep in mind, though, that the requirements for a planet to be considered habitable are pretty broad, the primary criterion being only that it can sustain liquid water, which can occur over a wide range of

---

* As far as I'm aware, he's not.

temperatures and conditions. Consequently, these planets are highly variable in all kinds of other attributes: temperature, atmospheric composition, radiation load, gravity, and geological composition, just to name a few.

We know that populations experiencing different conditions tend to evolve in dissimilar ways, even if exposed to the same general selective pressures. And that evidence comes from populations of *E. coli*, sticklebacks, and fruit flies experiencing only slightly different environments. Surely, the very divergent environments across diverse planets would guide evolution down alternative paths.

But enough hypotheticals. All we really need to do is compare New Zealand to anywhere else in the world to see contrasting evolutionary worlds and how they've unfolded. In the greater scheme of things, birds and mammals aren't all that different. They're not only carbon-based and use DNA, but both are vertebrate animals sharing many basic aspects of biological functioning. Yet the fauna of New Zealand is strikingly different from that of Australia, the Andes, the Serengeti, and everywhere else. No one would compare New Zealand to any of these places and say that evolution had occurred in a similar way.

Or compare the reign of the dinosaurs—*T. rex, Stegosaurus,* and one-hundred-foot-long, seventy-ton sauropods—with the world today, comprised of elephants, giraffes, cats, and krill-eating blue whales. Maybe a *Triceratops* is vaguely like a rhinoceros, and perhaps a *Struthiomimus* is similar to the ostrich it mimics, but for the most part, the dominant animals of the Mesozoic were vastly different from those that replaced them.

If life evolves so differently on this planet across space and time, it doesn't make much sense to expect that life on other planets will parallel what's evolved here on Earth. Carl Sagan put it well in his bestseller *Cosmos,* a half dozen years before Gould's *Wonderful Life:* "Some

Masters of the Mesozoic, with no parallel before or since

people—science fiction writers and artists, for instance—have speculated on what other beings might be like. I am skeptical about most of those extraterrestrial visions. They seem to me to rely too much on forms of life we already know. Any given organism is the way it is because of a long series of individually unlikely steps. I do not think life anywhere else would look very much like a reptile, or an insect or a human."

WHAT ABOUT THE IDEA that humans—or something like us—were destined to evolve? Conway Morris is not the only person to have made such a claim, but he is the most specific, suggesting that even if the asteroid had not hit, mammals still would have flourished and humans would have evolved once the Earth cooled thirty million years ago. But this claim, too, flies in the face of what actually happened during the course of evolution. If the evolution of humans is so inevitable, why did it happen only once?

Consider again the Empire of Birds, New Zealand, which has gone its own evolutionary way for the past eighty million years. Where are the humanoids there? The closest evolution has come to producing even a mammal is the little worm-snuffling kiwi. Australia, also on a geologically separate evolutionary trajectory, started with mammals, but the nearest it came to humans was the vaguely monkey-like tree kangaroo, known neither for its wits nor for any other human-like characteristic.

Even places seeded with primates failed to evolve humans. Lemurs, which floated over to Madagascar more than forty million years ago, are a great evolutionary success story, producing a wide range of species. But despite beginning to diversify tens of millions of years before the first hominids arose from our simian ancestors, lemurs have not produced any even slightly humanesque species. South America was isolated from the rest of the world for about fifty million years until the rise of the Isthmus of Panama a few million years ago. Monkeys washed ashore there, too, having drifted over from Africa on a log or other vegetation around thirty-six million years ago. They also became quite diverse—from tiny marmosets to gangly spider monkeys—but no humanoids evolved on that island continent, either.

The fact is, we humans are an evolutionary singleton—nothing else like us has ever evolved on Earth anywhere, any time. The ubiquity of convergent evolution in general would seem to provide scant support for our evolutionary inevitability.

LET'S CONSIDER EXTRATERRESTRIAL LIFE from another perspective. "Are we alone?" is one of humanity's great questions. The answer depends on what we mean by "we."

Does "we" refer to life on other planets? If so, obviously, we don't know the answer. But given those billions of habitable, Earth-like planets, many scientists consider it inevitable that life has evolved elsewhere, perhaps many times.

Assuming life has evolved on other planets, would it be multicellular and complex, like here, or just a bunch of simple, single-celled organisms? Life began on Earth almost four billion years ago. The first organisms were minute and composed of a single cell, and this condition persisted for two and a half billion years (give or take a few hundred million). Eventually, however, organisms composed of multiple cells evolved, and they did so many times. By conservative estimates, multicellularity arose minimally once in animals, three times in fungi, six times in algae (including land plants, which evolved from one type of green algae), and at least three times in bacteria. More liberal estimates—hinging on what counts as "multicellularity"—yield estimates of at least twenty-five origins.

Scientists also debate what is meant by the term "complexity"—an organism with more than one cell is not necessarily much more complicated than a single-celled creature. Proposed measures of complexity include the number of different types of cells in an organism (for example, we have muscle, skin, and many other types of cells), the number of different specialized parts, the number of different types of interactions among different parts, and many more. Regardless of which definition is used, complexity has clearly evolved many times among Earth's inhabitants. This great repeatability suggests that multicellularity and complexity are an inevitable outcome of the evolution of life, at least here on Earth. If life has evolved on other Earth-like planets, we earthlings may not be alone in our multicellular complexity.

We now know that intelligence (yet another term with many disparate definitions) has also evolved convergently many times, not just in our primate relatives, but in a wide variety of other animals. Elephants placing boxes in just the right spot to stand on them and reach food, crows making tools with hooks to snag grubs in crevices, dolphins using symbolic language to answer questions about whether an object is present in their tank—many large-brained species are much more clever than we previously realized. Even species never thought to have much in the way of smarts, like lizards and fish, are proving capable of performing challenging cognitive tasks. Perhaps the most surprising are octopuses, which, despite having brains structured very differently from ours, can figure out how to unscrew jars and use coconut husks as a disguise as they move across the ocean floor.

In addition to problem solving, at least some animals have self-awareness, usually tested by putting a mark on an animal and then placing a mirror in front of it. If the animal looks at the mirror and touches the mark, then it must have recognized that it is looking at itself. We used to think that only apes were capable of such self-recognition, but in recent years elephants, dolphins, and magpies have passed the mirror test. In fact, a number of these animals have used the mirrors to look into their mouths and at other body parts they can't normally see, clear evidence that animals are aware and curious about themselves.

So, advanced intellectual capabilities have arisen convergently many times on Earth. If that's any indication of how life may have evolved on other planets, we may not be the only intelligent beings in the universe.

Despite this convergent evolution of intelligence and self-awareness, no other animal has developed anything like our intelligence. Nor, as far as we're aware, have other animals evolved anything like our ability for self-conscious introspection. Given this single, non-convergent origin, we can't say whether the evolution of our intelligence was a highly unlikely fluke—Earth's own Cit+—or something that inevitably would have happened. As a result, we can't look to what's happened on Earth to predict whether comparable—or greater—levels of intelligence may have evolved on other planets.

LET'S RETURN TO EARTH. In a twist on Gould's metaphor, let's replay the tape, but with one major change: no us. If we hadn't evolved, can we predict whether highly intelligent, mindful beings would have evolved?

Certainly, the prerequisites—a large brain and some level of intelligence—are present in other primates. As we've seen, closely related species are prone to evolve in parallel. Absent humans, might apes or some other primate species have taken our place?

Paleoanthropologists have long debated what triggered the sudden and rapid increase in brain size in our lineage. The fossil record indicates that the evolution of bipedality preceded the increase in brain size. The occupation of open savannah habitats has also been implicated as a driving factor. Could another primate take the same path? It doesn't seem impossible. In fact, a population of chimpanzees in Senegal inhabits savannahs and uses spears to hunt smaller primates; they also occasionally walk bipedally, though anatomically they're not well adapted for it. Even without invoking *Planet of the Apes*, it's not hard to envision them evolving human levels of sentience.

What about non-primates? They're not closely related to humans and thus don't share our genetic proclivities in the way that Senegal chimpanzees and other primates do. Is it possible that elephants, dolphins, crows, or octopuses could evolve human-level intelligence? I can't say it's impossible. All of these types of animals have been around for many millions of years without any of them doing so, but given enough time, who knows?

And, finally, what about the dinosaur that would be human? Replaying the tape yet again, but with a different counterfactual, let's return to the premise of *The Good Dinosaur*, that asteroid whizzing by Earth, a near miss, a non-cataclysm. *Velociraptor, Troodon,* and friends survive with their big brains. Where does evolution take them?

Dale Russell postulated that natural selection would favor ever-increasing saurian brain size, leading to a series of anatomical transformations that ultimately would produce a green reptilian startlingly similar to you and me. Since Russell published that paper, Conway Morris and others have cited it in support of the proposition that a human-like species was destined to evolve.

We know now that Russell got one detail wrong. Back in the early 1980s when Russell published his paper, paleontologists were still debating whether birds evolved from dinosaurs. Now that dispute is settled—all but a few mavericks agree that birds descended from a group of theropods closely related to *Velociraptor* and *Troodon*. Not only are birds now recognized as a branch of the dinosaur evolutionary tree, but new findings have revealed something not even hinted at thirty-five years ago. Thanks to spectacular fossil discoveries in China, we now know that feathers evolved early in theropod dinosaur history, before one type of theropods gave rise to birds. As a result, not just birds, but many theropods—including even baby *T. rex*es, as well as *Troodon*—were covered in feathers.*

Consequently, Russell's depiction requires some updating. Instead of green, scaly skin, we need to clothe Mr. Dinosauroid in a cloak of feathers. And while we're at it, let's give him some jaunty coloration, like a parrot. No longer will anyone mistake this handsome fellow for an extraterrestrial!

More important than the external appearance, however, was the series of transformations proposed by Russell that took the horizontally oriented *Troodon*—head and forebody pitched forward, counterbalanced by a long tail, the body precisely balanced over its hindlimbs—

---

* Technically, fossil feathers of neither *Troodon* nor young *T. rex* are known, but in both cases, fossils have been found of very closely related species bearing feathers.

and transformed it into a vertical biped. Would the evolution of a bigger brain really have required this?

Russell didn't provide a lot of detailed explanation for why he thought that the evolution of a larger braincase necessarily requires an upright body posture. Even the brainiest of birds—and some are quite smart, with very large brains for their body size—continue in their horizontal, dinosaurian body posture. And *T. rex*, with its enormous skull, didn't walk erect.

Russell's critics contend that his perspective was too anthropocentric, that his predictions were too guided, at least subconsciously, by how humans evolved. "Much too human," said one paleontologist; "suspiciously human," echoed another.

Russell expected this objection and countered it in his paper. Citing convergent evolution (and anticipating Conway Morris by more than a decade), he argued that if humans evolved one time, then certainly it's reasonable to expect something like us to evolve again ("If it's such a good solution for us, is it so difficult to imagine it could be a good solution for a dinosaur?" Conway Morris later agreed). Still, Russell acknowledged that his ideas were speculative, and concluded his paper by throwing down the gauntlet: "We invite our colleagues to identify alternate solutions."

The gauntlet was not picked up. In fact, Russell's paper has not been widely discussed in the scientific literature. Since its publication in 1982, it has only been referenced forty-one times by other scientific papers—that's about one per year, scarcely a citation classic. Moreover, most of the references were in hard-core paleontology papers, discussing the technical details of *Troodon*'s anatomy (the primary focus of Russell's paper); the dinosauroid idea received little scientific attention or response.

I've known about the paper since the late 1980s and considered it to be an undiscovered gem that I was looking forward to bringing to the light of day in this book. But then one day it occurred to me that perhaps I should check the internet to see if there was any mention of Russell's idea online. And there, much to my surprise, I discovered a vast array of blog posts, comment threads, illustrations, film clips, and interviews. The dinosauroid hypothesis was alive and thriving in cyberspace.

This online material can be split into three groups. First are the uncritical reports of Russell's idea, simply repeating what he wrote, usually accompanied by drawings or photographs of the sculpture he produced.

The second group revolved around the alien-like appearance of Russell's reconstruction. Green-skinned, scaly, human-like, with slightly odd features and missing a few parts—the dinosauroid looks the epitome of a science fiction extraterrestrial. And that, in turn, led to all manner of wacky discussion, exemplified by a website reporting (apparently seriously) that the dinosauroid "evolved into a humanoid species that eventually developed a culture that ran its course or was destroyed in an Atlantis-like catastrophe—just after they had begun exploring extraterrestrial frontiers. Certain UFOnauts, then, may be the descendants of the survivors of that reptile culture RETURNING from their space colony to monitor the present dominant species on the HOME planet."

Amidst all this nonsense, the third group of responses has come from a number of paleontologists and dinosaur enthusiasts that have taken the dinosauroid seriously, critiquing the hypothesis and providing the alternative scenarios Russell invited. These authors make a familiar point: species that are initially different will not take the same evolutionary route in response to similar selective pressures.

Consider how we got to where we are. Our closest relatives are the apes—gorillas, chimps, orangutans, and gibbons. None of these species has a tail. You have to go deep in our evolutionary history, to the point where our ape-human lineage diverged from Old World monkeys about twenty-two million years ago, to find a tailed ancestor. Only well after tails were lost did our ancestors begin walking on two legs. Consequently, it's no surprise that we evolved a completely erect posture to balance our big heads directly over our bodies—we didn't have any posterior appendage that could serve as a counterweight. If we hadn't become completely erect, we'd always be pitching forward, struggling to keep our balance.

Now picture a pigeon walking on the sidewalk. It's bipedal, too, but its head is held out in front and its body is horizontal, not vertical. The legs are positioned like a seesaw's fulcrum, placed just right so that the bird topples neither forward nor backward. Scale that pigeon up to about three feet in height, put teeth in its jaws, swap out the wings for arms, make a few other tweaks, and you've got a *Troodon*.

Let's suppose that natural selection favors bigger brains, and thus a bigger head, in *Troodon*. For a bipedal animal lacking a tail, an erect posture is clearly the most effective. But for a species that already uses a tail to balance the front end of the body, an easier evolutionary route would be to build a heavier tail.

That's not to say that Dale Russell's scenario isn't possible—evolution conceivably could have proceeded in that direction. But there's no reason to expect that a scenario derived from how tailless apes turned into humans would apply to a very different ancestor with different anatomical features.

Taking this perspective, several alternatives to the dinosauroid have been proposed. Although varying in detail, they all share a core similarity: the sci-fi reptilian-alien look is out and in its place is a large, big-brained, bird-like animal. Adorned with feathers, perhaps using a beak in addition to or in place of hands to manipulate objects, these

A new view of the dinosauroid

highly intelligent dinosaur descendants strut about like an egret or crow, body horizontal to the ground, tail balancing its large head, tools in hand or beak.

There's no way of knowing whether evolution might have gone down this path. But one conclusion seems clear: if the dinosaurs had survived, it's not obvious that their descendants today—even the really smart dinosaurs—would be at all similar to us. A super-sized brainy chicken would probably be closer to the mark.

RECENTLY I BECAME AWARE of one of the most unlikely secret agents. Referred to in his catchy theme song as "a semi-aquatic egg-laying mammal of action," Perry the Platypus thwarts evildoers in their efforts to take over the tri-state area.* Teal blue with an orange bill and feet,

---

* Never explicitly defined, but apparently in the vicinity of Denver.

clad in his trademark brown fedora, Perry is a Down Under 007: charming, witty, and sporting advanced jujitsu skills, he possesses an armory of technical gizmos and pilots a flying platypus-mobile.

Admittedly, *Phineas and Ferb*, which ran for four seasons on the Disney Channel, took a few biological liberties with its furry star. Adult platypuses, after all, do not have teeth, nor do they walk bipedally or make a chattering, growly noise that drives human females wild with passion. Still, the program has brought well-deserved recognition of this wonderful species to a generation of children, even extolling some of the platypus's many adaptations, including its thick fur, powerful tail, webbed feet, and, of course, its bill. Indeed, labeled as "primitive" because it is one of only a few mammals to lay eggs, the platypus often doesn't get its due. As I watched the show, I came to realize that Perry ought to be very proud of himself and his kind.

That led me to ponder what other thoughts might go through Perry's crime-fighting mind. From my limited perusal of Perry's oeuvre, it's not clear how introspective he is. But assuming that he does take time away from counter-espionage to think deep thoughts, one rumination might be about the plausibility of platypusoid life on other planets. And why not? Duck-billed platypuses are extraordinary animals, from Perry's perspective no doubt the pinnacle of the evolutionary process. It would be only natural for a pensive platypus to wonder if *Ornithorhynchus anatinus* is alone in the cosmos.

Some might object that this is ridiculous. Platypus at the peak? With their small brains and lack of language? No, they might cavil, platypuses are only a sideshow, a way station along the road to the apogee of evolution, us. After all, we truly are the zenith, with our big brains and tools and consciousness and so on.

But, of course, that view is highly anthropocentric. We have our good points, don't get me wrong. But so, too, does the duckbill.

Indeed, let's talk about that bill. Superficially similar to the beak of a duck (hence the name), the platypus' leathery rostrum is covered with tens of thousands of minute sensors. Sixty thousand of them are extraordinarily sensitive to touch, able to detect changes in water pressure produced by the flick of a fish's fin. Another forty thousand, however, have a different function.

How the duckbill finds its food was long a mystery. When a platypus enters water, it shuts its eyes, ears, and mouth. Yet, despite this sensory deprivation, it is able to catch and eat half its weight in crayfish every night. It does spend some time rooting around the streambed with its bill, much like a duck does, but this is not an efficient way to find prey. A hint comes from the way the platypus swims through the water, its head and bill swinging back and forth nonstop. These observations led the Australian naturalist Harry Burrell to hypothesize early in the twentieth century that the platypus must have some sort of sixth sense.

The nature of this additional faculty was revealed in the 1980s by a team of Germans and Australians who discovered that a platypus could find a battery emitting a weak electric charge even when it was hidden on the bottom of a pool. Further work has confirmed that thanks to the electroreceptors on their bill, platypuses are capable of precisely locating food by the minute electrical charges their prey produce as they move. Combined with their ability to detect water currents and other movements with their touch receptors, platypuses essentially "see" their world in terms of tactile and electric stimuli.

So platypuses are just as extraordinary in their own way as we are in ours. And thus why not ask if platypusoid life has arisen elsewhere?

Unfortunately, the case for platypusoids on other planets is just as uncompelling as that for extraterrestrial humanoids. The platypus, too, is an evolutionary singleton, one that only evolved in Australia.*

---

* Several sixty-million-year-old molar teeth from Argentina suggest that ancient platypus relatives were once present in South America, probably a result of the ancient Gondwanan connection between the two continents.

The failure of other mammals to evolve into platypus-like creatures on other continents, despite the occurrence of similar environments, makes it hard to claim that the platypus way of life represents the supreme form of adaptation to stream environments, that to which natural selection must inevitably strive. As a result, though intergalactic platypuses would be wonderful, I see no reason to put them on the top of our list of expected extraterrestrials. The platypus, like humans, gets placed firmly in the contingency category, an unparalleled species that evolved in one place and nowhere else.

But even though evolution has produced no platypus doppelgängers, Perry's kind may be the crown jewel of convergent evolution, each platypus part paralleled in some other organism. Who can blame early English scientists for thinking it was a hoax, a craftily assembled amalgamation from disparate beasts? And, indeed, it is: bill of a duck; webbed feet of an otter; lush, watertight fur of a sea otter;* stout tail somewhat like that of a beaver. Even its electroreception capabilities are paralleled in electric eels and other fish, the Guiana dolphin, and a type of salamander. And although no other animal has a venom-injecting spine on its ankle like that of the platypus, the anatomy of this structure is convergent with a rattlesnake's fang: both are hollow tubes attached to a venom gland with muscles that squeeze the gland, pushing the venom through the spur or fang and into its target. So, the platypus is paradoxically both a paragon and a repudiation of convergence, evolutionarily unique, but a composite of convergent traits.

The duckbill is not alone in this ambiguity. We humans, too, are an evolutionary singleton possessing many features independently evolved in other lineages, including:

---

* So dense that platypuses can swim in near-freezing water while barely losing any body heat.

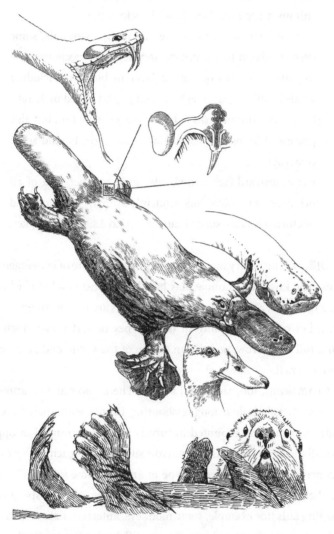

*The platypus, an evolutionary singleton that is a mélange
of parts convergent with other species*

- Bipedality, shared with birds and their theropod dinosaur relatives, kangaroos, and hopping rodents, not to mention the many animals that are occasionally bipedal, such as chimps, pangolins, lizards, and cockroaches.
- Reduced hair, a trait exhibited by many mammals, some even depilated to a much greater degree, especially those in warm climates or with a layer of blubber, including whales, hippos, pigs, elephants, seals, and naked mole rats.
- Opposable thumbs, shared with our primate kin, but also possessed by opossums, koalas, some rodents, and some tree frogs.
- Large, forward-facing, binocular eyes, another trait of all primates, but evolved independently by many predatory and nocturnal species, such as cats, owls, and Asian whip snakes.

Other evolutionary one-offs have their own share of convergent features. The kiwi, for example, has hair-like feathers and stiff whiskers like a mammal; a number of its other features are rare or unique among birds, but occur commonly in other types of vertebrates, including bones full of marrow, nostrils on the tip of the snout, and an excellent sense of smell.

Chameleons, too, are one of a kind. There's no other creature like these prehensile-tailed, tongue-shooting lizards with their independently rotating eyes mounted in turrets and toes arranged in opposition, all the better for grasping narrow surfaces. But each of these traits is convergent with a different type of animal: some salamanders also fire their tongues far out of their mouths; many arboreal species have grasping tails (for example, some monkeys, anteaters, lizards, possums, and seahorses); the sand lance, a type of fish, also has turreted and independently moving eyes; and some birds and marsupials have moderately similar toe arrangements.

So maybe we're selling convergence short. We evolutionary singletons—kiwis, platypuses, chameleons, and humans—may be unique, but many of our parts have evolved convergently in other organisms.

In other words, here on Earth, species frequently do evolve similar features in response to similar environmental conditions. So, even if a humanoid or a platypusoid (or a chameleonoid or kiwioid) is unlikely to have evolved elsewhere, that's not to say that extraterrestrials would look completely unfamiliar. An extraterrestrial might even be a mashup, platypus-style, of many different parts borrowed from different Earth inhabitants.

THE GREAT EVOLUTIONARY BIOLOGIST Edward O. Wilson recently speculated about the biology of extraterrestrials capable of developing a civilization as advanced as ours. His predictions, based on what we know about evolution on Earth, are that they would:

- Be terrestrial because the development of technology would have required harnessing a transportable form of energy, such as fire.
- Be large because large brains are needed for the neural processing necessary for great intelligence.
- Rely on visual and auditory systems for communication because these are the most effective ways of transmitting signals over long distances.
- Have a large head mounted at the front of the body. The head would be big to contain the large brains, as well as the sensory organs used to survey surroundings as the creatures moved forward.
- Have jaws and teeth for procuring prey, but not overly large ones. The social cooperation required for a species to

develop advanced civilization would guarantee that prey
capture and defense were accomplished through coopera-
tive means and intelligence, rather than simply by brute
force.

• Possess a small number of limbs or other appendages, with
at least one pair with pulpy tips good for sensitive touch
and manipulation.

Although these prescriptions are a lot less specific than Dale Russell's, many commenters probably would level the same criticism—that
the list is still too parochial, too constrained by what actually has
evolved here on Earth. In Wilson's defense, that was explicitly his starting point, his deductions drawn from Earth's evolutionary history.

But let's put these criticisms aside and take Wilson's suggestions at
face value—they certainly represent reasonable predictions about what
a technologically adept extraterrestrial species might be like. And these
predictions really aren't very restrictive. In this framework, the extraterrestrial species might be a two-legged biped with a large head, two
eyes, a small mouth with small teeth, a pair of appendages on either
side of its upper body, and fine filaments on top of its round head. Or it
could have eight jointed appendages, six used for moving and the front
two with seven more delicate jointed appendages at the tip, six enormous saucer-shaped openings to detect sound across its bulbous head,
with a rotating stalk attached to the top with three enormous eyes
arranged as a triangle. In other words, even with Wilson's terracentric
view of extraterrestrial evolution, the end result could be either very
similar to what occurs on Earth or completely different, with parts that
are only vaguely similar to those owned by Earth species.

And that brings me to a more general point. Should we really be
looking at life here on Earth to predict what it might be like else-

where? I, for one, am not at all convinced that life on Earth has uncovered every conceivable way of existing on a planet like our own, or even most of the ways.

Is there any reason that a plant-like organism—one that gets its energy straight from its sun or perhaps through some chemical reaction—couldn't be mobile, evolving limbs or some other way of moving around? And if they did so, wouldn't that require a nervous system that could ultimately lead to the evolution of intelligence?

And who says that limbs are necessary for getting around? Octopuses and squid move by jet propulsion, shooting water out of a tube to propel themselves in the opposite direction. In some atmospheres, that might be very effective.

Let's recall those weird life-forms populating the Burgess Shale. Oozing half-pinecones, predators with five eyes and a claw on a hose hanging off its front end, worm-like tubes on squiggly legs with rows of spines on their backs, a floating Band-Aid with a mouth on the underside. These animals actually existed. Who's to say that something like them couldn't be the ancestors of modern ecosystems on other planets? And if so, what would life on those planets look like today?

AT THE END OF THE DAY, we know that evolution is not random or haphazard. Natural selection restricts the way that species can evolve, often constraining them to adapt in the same way when facing similar environmental circumstances. In some cases, there are single best biological solutions to problems posed by the environment, and in many cases, species repeatedly attain these optima. Moreover, related species share many similarities in all aspects of their biology—their genetic similarities and the way they develop being particularly important. These similarities, too, bias close relatives to follow the same evolutionary course. Consequently, convergent evolution usually results from the conjunction of a limited number of optimal solutions and shared similarities in genetics, development, and ecology funneling adaptation in the same direction.

The world of biological possibilities, however, is often a vast one, and even with biases from natural selection, genetics, and development, the set of evolutionarily realizable end points may be large. As a result, evolution often goes its own way. This is particularly true when evolution begins from different starting points with different genes and developmental systems. However, even starting from the same ancestral stock and experiencing similar circumstances, the outcome can be divergent. Evolution repeats itself sometimes, but often it doesn't.

So, can we predict evolution? In the short-term, yes, to some extent. But the longer the passage of time and the more different the ancestors or conditions, the less likely we are to prognosticate successfully. Dinosauroid? I don't think so. Perry the Platypusoid? Alas, no. Were we destined to be here? Hardly.

If any of a countless number of events had occurred differently in the past, *Homo sapiens* wouldn't have evolved. We were far from inevitable and are lucky to be here, fortunate that events happened just as they did. Asteroids, of course, but what other events critically tipped evolution's path in our favor? Who knows how slight a difference in the past—a tree falling on great-great-great-to-the-millionth-degree-grandpa Ernie, a forest fire, a mutation—might have snuffed out our future existence?

On the other hand, perhaps with a different historical sequence, humanoid doppelgängers could have evolved prolifically. Perhaps the world could have been populated by marsupial humans, as well as lemur humans, bear humans, crow humans, even lizard humans. Imagine the United Nations, but with each representative a different evolutionary lineage. It could have been.

From the vantage point of the origin of life several billion years ago, any particular evolutionary outcome would have seemed improbable. But history happened as it happened and here we are today, the result of billions of years of natural selection and the flukes of history that sent life down one path and not others. Lucky? Yes. Destined, no. We should make the most of our evolutionary good fortune.

Acknowledgments

I am deeply grateful to a veritable city of friends, colleagues, and relatives who provided help during the three years I worked on this book. Many people answered questions small and large. I'm particularly grateful to those whom I badgered seemingly endlessly, even subjecting them to reviewing what I'd written, including Rowan Barrett, Zack Blount, Fred Cohan, Tim Cooper, John Endler, Marc Johnson, Rees Kassen, Craig MacLean, Rasmus Marvig, Paul Rainey, David Reznick, Dolph Schluter, Roy Snaydon, Bruce Tabashnik, Michael Travisano, and Nash Turley. Thanks, too, to the many others who provided information, advice, and answers: Eldridge Adams, Anurag Agrawal, Chris Hamlin Andrus, Spencer Barrett, Dan Blackburn, Chris Borland, Angus Buckling, Molly Burke, Todd Campbell, Scott Carroll, Gary Carvalho, Satoshi Chiba, Ke Dong, Mick Crawley, Stuart Davies, Chuck Davis, Doug Erwin, Scott Edwards, Maha Farhat, Charles Fox, Gonzalo Giribet, Pedro Gómez López, Billie Gould, Wendy Hall, Chris Hamlin, Marshal Hedin, Andrew Hendry, David Hillis, Hopi Hoekstra, Nina Jablonski, George Johnson, Leo Joseph, Betul Kacar, Rick Lankau, Tami Lieberman, Adrian Lister, Tim Low, Zhe-Xi Luo, Andy Macdonald, Blue Magruder, Jordan Mallon, Greg Mayer, Axel Meyer, Mark Moffett, Loretta O'Brien, Mark Olson, Sterling Nesbitt, Mike Palmer, Delphine Picard, Gregory Priebe, Peter Raven, Diana Rennison, Robert Ricklefs, Sara Ruane, Eric Rubin, Dov Sax, Tom Schoener, Phil Service, Susan Singer, Russell Slater, Morten Sommer, David Spiller, Jonathan Storkey, Yoel Stuart, Doug Swain, Corina Tarnita, Henrique Teotonio, Erdal Toprak, Ken Weber, and Andrew Whitehead. In addition, in a number of cases, I posted on Facebook asking for examples of this or that and received scores of great suggestions from a vast number of FB friends—thanks to all of you who responded. Many thanks to the library staffs at Harvard's Ernst Mayr Library (Ronnie Broadfoot, Connie Rinauldo, Dorothy Barr, and especially Mary Sears) and Washington University's Olin Library for help in locating hard-to-find references. Also many thanks to Jared Hughes for all sorts of help. The most effusive thanks go to those who read drafts of multiple chapters or even the entire book. That was going above and beyond! Thank you, Alan Barker, Frank Grady, Harry Greene, Wendy Hall, Ambika Kamath, Andy Knoll, Carolyn Losos, Joseph Losos, Ann Mandelstamm, Marc Mangel, Irwin Shapiro, and Mike Whitlock. Thanks, too, to David Reznick and Cody Lane for facilitating my visit to Trinidad.

Neil Shubin, Dan Lieberman, Nicholas Dawidoff, and especially Doug Emlen provided invaluable advice about the book-writing process. Max Brockman, my literary agent, was wonderful in getting the process going, and Courtney Young, my editor, did a fabulous job in taking what I had written and turning it into something much better. Thanks, too, to Kevin Murphy, Alexandra Guillen, Martha Cameron, Joel Breuklander, and the rest of the Riverhead Books team for help turning my manuscript into a book. Doug Tuss and Emily Harrington did a wonderful job in laying the groundwork for the illustration program, and Marlin Peterson's illustrations are superb and were brilliantly executed in a short window of time.

Finally, thanks to my parents, Carolyn and Joseph Losos, for all of their support throughout my life, but especially for their enthusiasm for this project, and to my beloved wife, Melissa, for advice and help in so many different ways, for never seeming bored as I prattled on about the latest fact I'd learned and for generally putting up with me during the entire process.

About the Illustrator

Marlin is an illustrator with a fanaticism for ping pong and all things zoological. He loves animals from the Cenozoic, and knows arachnids don't get the respect they deserve. He is an art instructor at Wenatchee Valley College where he teaches courses such as drawing geology and science illustration. When not dabbling in his garden, he can be found freelancing commissions, and plotting trompe l'oeil murals. He is fluent in many forms of traditional and digital media, and finds ways to dabble in them all.

Visit his website to see more of his projects.
marlinpeterson.com

Improbable Destinies was exhilarating to illustrate thanks to the broad swath of subjects. Marlin thoroughly enjoyed researching and drawing the many animals he illustrated for this book, and thanks Jonathan for his keen eye for detail, great feedback, and amazing vision for this book.

Gratitude leaking out of every pore,
deepest love for Christine and Chess.

Notes

In this section, I provide references and additional information on a few points. The references are not meant to be exhaustive; rather, for many topics I have provided one or two papers that provide an entrée to a topic. Many of the examples discussed in the text (for example, the evolution of guppies, anoles, and the peppered moth, as well as Rich Lenski's Long-Term Evolution Experiment) are discussed in great detail in entries on the internet. I have provided references to specific scientific papers discussed in the text, as well as those papers that are directly quoted.

introduction: THE GOOD DINOSAUR

3 **long-necked brontosaurs:** E. Tschopp, O. Mateus, and R. B. J. Benson. 2015. A specimen-level phylogenetic analysis and taxonomic revision of Diplodocidae (Dinosauria, Sauropoda). *Peer J* 3: e857.

5 **"the rise of . . . ape-like mammals":** p. 222 in S. Conway Morris. 2003. *Life's Solution: Inevitable Humans in a Lonely Universe.* Cambridge, UK: Cambridge University Press.

6 **Canadian paleontologist Dale Russell:** D. A. Russell and R. Séguin. 1982. Reconstructions of the small Cretaceous theropod *Stenonychosaurus inequalis* and a hypothetical dinosauroid. *Syllogeus* 37: 1–43.

8 **Morris even appeared in a BBC documentary:** "My Pet Dinosaur," an episode on the BBC show *Horizons,* which aired March 13, 2007. https://www.youtube.com/watch?v= rmaLa_6o_Qg.

8 **four light-years away:** D. Overbye. 2013. Far-off planets like the Earth dot the Galaxy, *New York Times,* November 4, 2013; Proximate goals, *Economist,* August 27, 2016, A1.

9 **"If we ever succeed in communicating":** p. 457 in R. Bieri. 1964. Huminoids on other planets? *American Scientist* 52: 452–458.

9 **David Grinspoon:** pp. 272–273 in D. Grinspoon. 2003. *Lonely Planets: The Natural Philosophy of Alien Life.* New York: HarperCollins.

9 **Not surprisingly, Conway Morris agrees:** p. 328 in Conway Morris. 2003. *Life's Solution* (see Introduction, n. 5).

13 **genetic studies starting in the 1980s:** C. G. Sibley and J. E. Ahlquist. 1990. *Phylogeny and Classification of Birds: A Study in Molecular Evolution.* New Haven, CT: Yale University Press; F. K. Barker et al. 2004. Phylogeny and diversification of the largest avian radiation. *Proceedings of the National Academy of Sciences of the United States of America* 101: 11040–11045.

15 **Taking this a step further:** p. 272 in G. McGhee. 2011. *Convergent Evolution: Limited Forms Most Beautiful.* Cambridge, MA: MIT Press.

15 **Conway Morris agrees:** P. Gallagher. 2015. Forget little green men—aliens will look like humans, says Cambridge University evolution expert. *The Independent,* July 1, 2015, http://www.independent.co.uk/news/science/forget-little-green-men-aliens-will-look-like-humans-says-cambridge-university-evolution-expert-10358164.html.

17 **Hitting home:** p. 289 in S. J. Gould. 1989. *Wonderful Life: The Burgess Shale and the Nature of History.* New York: W. W. Norton.

18 **However, in an exchange:** S. Conway Morris and S. J. Gould. 1998. Showdown on the Burgess Shale. *Natural History:* 107(10): 48–55.

18 **Nothing like it exists:** M. Henneberg, K. M. Lambert, and C. M. Leigh. 1997. Fingerprint homoplasy: koalas and humans. *Natural Science* 1: 4.

chapter one: EVOLUTIONARY DÉJÀ VU

28 **In 2013, a team of Sri Lankan:** K. D. B. Ukuwela et al. 2013. Molecular evidence that the deadliest sea snake *Enhydrina schistosa* (Elapidae: Hydrophiinae) con-

sists of two convergent species. *Molecular Phylogenetics and Evolution* 66: 262–269.

30 **In a paper published in 2014:** F. Denoeud et al. 2014. The coffee genome provides insight into the convergent evolution of caffeine biosynthesis. *Science* 345: 1181–1184.

38 **because we wouldn't be here:** D. H. Erwin. 2016. *Wonderful Life* revisited: chance and contingency in the Ediacaran-Cambrian radiation. Pp. 277–298 in G. Ramsey and C. H. Pence, eds., *Chance in Evolution.* Chicago: University of Chicago Press.

38 **as Gould repeatedly, exaltingly, emphasized:** Both quotes from p. 572 in S. Conway Morris. 1985. The Middle Cambrian metazoan *Wiwaxia corrugata* (Matthew) from the Burgess Shale and *Ogygopsis* Shale, British Columbia, Canada. *Philosophical Transactions of the Royal Society of London B* 307: 507–582.

39 **To some extent:** Erwin, *Wonderful Life* revisited, 277–298 (see Chapter One, n. 38).

39 **In addition, some analyses have compared:** Technically, this analysis was restricted to arthropods, which are the invertebrates with jointed legs, such as spiders, lobsters, and insects. Many—but not all—of the most interesting Burgess Shale species are arthropods. D. E. G. Briggs, R. A. Fortey, and M. A. Wills. 1992. Morphological disparity in the Cambrian. *Science* 256: 1670–1673; M. Foote and S. J. Gould. 1992. Cambrian and recent morphological disparity. *Science* 258: 1816.

40 **One colleague proposed that Gould's views:** P. Bowler. 1998. Cambrian conflict: crucible an assault on Gould's Burgess Shale interpretation. *American Scientist* 86: 472–475.

40 **Another suggested that Conway Morris:** R. Fortey. 1998. Shock Lobsters. *London Review of Books,* October 1, 1998, 24–25.

41 **In our conversation:** These thoughts appeared in Conway Morris' review of Gould's *Bully for Brontosaurus.* S. Conway Morris. 1991. Rerunning the tape. *Times Literary Supplement* 4628 (London), December 13, 1991: 6.

41 **"Evolutionary convergence is completely ubiquitous":** Gallagher. Forget little green men (see Introduction, n. 15).

46 **Says Conway Morris:** Ibid.

46 **Convergence extends even to the placenta:** D. G. Blackburn and A. F. Fleming. 2012. Invasive implantation and intimate placental associations in a placentotrophic African lizard, *Trachylepis ivensi* (Scincidae). *Journal of Morphology* 273: 137–159.

51 **What makes this case of convergence:** S. Conway Morris. 2014. *The Runes of Evolution: How the Universe Became Self-Aware.* West Conshohocken, PA: Templeton Press.

52 **The adaptive significance of variation:** For more on skin-color evolution, see N. G. Jablonski. 2012. *Living Color: The Biological and Social Meaning of Skin Color.* Berkeley: University of California Press.

53 **With cows:** Got lactase? *Understanding Evolution,* April 2007, http://evolution.berkeley.edu/evolibrary/news/070401_lactose.

chapter two: REPLICATED REPTILES

58 **Jamaica, a tenth:** For more information on anoles, see my previous book: J. B. Losos. 2009. *Lizards in an Evolutionary Tree: Ecology and Adaptive Radiation of Anoles.* Berkeley: University of California Press.

62 **a new means of closing wounds:** S. Reinberg. 2008. Gecko's stickiness inspires new surgical bandage. *Washington Post,* February 19, 2008, http://www.washingtonpost.com/wp-dyn/content/article/2008/02/19/AR2008021901653.html.

65 **Over time:** For example: J. A. Coyne and H. A. Orr. 2004. *Speciation.* Sunderland, MA: Sinauer Associates; P. A. Nosil. 2012. *Ecological Speciation.* Oxford, UK: Oxford University Press.

68 **convergence of entire radiations:** J. B. Losos. 2010. Adaptive radiation, ecological opportunity, and evolutionary determinism. *American Naturalist* 175: 623–639.

68 **little-known Japanese islands:** S. Chiba. 2004. Ecological and morphological patterns in communities of land snails of the genus *Mandarina* from the Bonin Islands. *Journal of Evolutionary Biology* 17: 131–143.

70 **Then DNA comparisons turned:** M. Ruedi and F. Mayer. 2001. Molecular systematics of bats of the genus *Myotis* (Vespertilionidae) suggests deterministic ecomorphological convergences. *Molecular Phylogenetics and Evolution* 21: 436–448.

71 **those in India:** F. Bossuyt and M. C. Milinkovitch. 2000. Convergent adaptive radiations in Madagascan and Asian ranid frogs reveal covariation between larval and adult traits. *Proceedings of the National Academy of Sciences of the United States of America* 97: 6585–6590.

71 **Despite great similarity:** S. Reddy et al. 2012. Diversification and the adaptive radiation of the vangas of Madagascar. *Proceedings of the Royal Society of London B* 279: 2062–71.

73 **biologists returning to islands:** For more information on island evolution, see: S. Carlquist. 1965. *Island Life: A Natural History of the Islands of the World.* Garden City, NJ: Natural History Press; R. J. Whittaker and J. M. Fernández-Palacios. 2007. *Island Biogeography: Ecology, Evolution, and Conservation,* 2nd. ed. Oxford, UK: Oxford University Press.

75 **Put any large mammal:** Ibid.

77 **I'll mention one last general trend:** A. S. Wilkins, R. W. Wrangham and W. T. Fitch. 2014. The "domestication syndrome" in mammals: a unified explanation based on neural crest cell behavior and genetics. *Genetics* 197: 795–808.

78 **experiment in Siberia:** L. Trut, I. Oskina, and A. Kharlamova. 2009. Animal evolution during domestication: the domesticated fox as a model. *BioEssays* 31: 349–360; L. A. Dugatkin and L. Trut. 2017. *How to Tame a Fox (and Build a*

Dog): Visionary Scientists and a Siberian Tale of Jump-Started Evolution. Chicago, IL: University of Chicago Press.

chapter three: EVOLUTIONARY IDIOSYNCRASY

83 **"the bat family's attempt to produce a mouse.":** p. 4 in J. Diamond. 1990. New Zealand as an archipelago: an international perspective. Pp. 3–8 in D. R. Towns, C. H. Daugherty, and I. A. E. Atkinson, eds., *Ecological Restoration of New Zealand Islands.* Wellington, NZ: New Zealand Department of Conservation.

85 **"resembling no other plant":** p. 208 in Carlquist, *Island Life* (see Chapter Two, n. 73).

95 **The French scientist François Jacob:** F. Jacob. 1977. Evolution and Tinkering. *Science* 196: 1161–1166.

97 **Not all cases of convergence:** T. J. Ord and T. C. Summers. 2015. Repeated evolution and the impact of evolutionary history on adaptation. *BMC Evolutionary Biology* 15: 137.

97 **The three-spined stickleback:** For more on sticklebacks and their evolution, see D. Schluter and J. D. McPhail. 1992. Ecological character displacement and speciation in sticklebacks. *American Naturalist* 140: 85–108; D. Schluter. 2010. Resource competition and coevolution in sticklebacks. *Evolution Education and Outreach* 3: 54–61; A. P. Hendry et al. 2013. Stickleback research: the now and the next. *Evolutionary Ecology Research* 15: 111–141.

99 **the number of toes:** D. B. Wake. 1991. Homoplasy—the result of natural selection, or evidence of design limitations? *American Naturalist* 138: 543–567.

100 **Ideally, we would directly test the hypothesis that natural selection has guided convergence:** How to study the role of natural selection in adaptation

and evolutionary convergence is discussed in A. Larson and J. B. Losos. 1996. Phylogenetic systematics of adaptation. Pp. 187–220 in M. R. Rose and G. V. Lauder, eds. *Adaptation.* San Diego: Academic Press, 187–220; and in K. Autumn, M. J. Ryan and D. B. Wake. 2002. Integrating historical and mechanistic biology enhances the study of adaptation. *Quarterly Review of Biology* 77: 383–408.

100 **"obviously there was some adaptive advantage":** Quoted in N. St. Fleur. 2016. Armed and dangerous: T-rex not the only dinosaur short-arming it. *New York Times,* July 19, 2016: D2.

chapter four: THE NOT-SO-GLACIAL PACE OF EVOLUTIONARY CHANGE

114 **Bernard Kettlewell:** H. B. D. Kettlewell. 1973. *The Evolution of Melanism: The Study of a Recurring Necessity with Special Reference to Industrial Melanism in the Lepidoptera.* Oxford, UK: Oxford University Press.

114 **The experimental demonstration:** L. M. Cook et al. 2012. Selective bird predation on the peppered moth: the last experiment of Michael Majerus. *Biology Letters,* February 8, 2012, DOI: 10.1098/rsbl.2011.1136.

117 **Rosemary and Peter Grant:** P. R. Grant and B. R. Grant. 1995. *40 Years of Evolution: Darwin's Finches on Daphne Major Island.* Princeton, NJ: Princeton University Press; J. Weiner. 1995. *The Beak of the Finch: A Story of Evolution in Our Time.* New York: Vintage Books.

121 **His much-debated theory:** S. J. Gould. 2002. *The Structure of Evolutionary Theory.* Cambridge, MA: Harvard University Press.

chapter five: COLORFUL TRINIDAD

126 **made the guppy an evolution celebrity:**
The following references provide a good
entrée to the more than half a century of
research on Trinidadian guppies: C. P.
Haskins et al. 1961. Polymorphism and
population structure in *Lebistes reticu-
latus*, an ecological study. Pp. 320–394
in W. F. Blair, ed., *Vertebrate Speciation.*
Austin: University of Texas Press; J. A.
Endler. 1980. Natural selection on color
patterns in *Poecilia reticulata. Evolution*
34: 76–91; D. Reznick and J. A. Endler.
1982. The impact of predation on life
history evolution in Trinidadian guppies (*Poecilia reticulata*). *Evolution* 36:
160–177; D. Reznick. 2009. Guppies and
the empirical study of adaptation.
Pp. 205–232 in J. B. Losos, ed., *In the
Light of Evolution: Essays from the Laboratory and Field.* Greenwood Village,
CO: Roberts & Co; A. E. Magurran.
2005. *Evolutionary Ecology: The Trinidadian Guppy.* Oxford, UK: Oxford
University Press; N. Karim et al. 2007.
This is not déjà vu all over again: male
guppy colour in a new experimental
introduction. *Journal of Evolutionary
Biology* 20: 1339–1350; D. J. Kemp et al.
2009. Predicting the direction of orna-
ment evolution in Trinidadian guppies
(*Poecilia reticulata*). *Proceedings of
the Royal Society of London B* 276:
4335–4343.

127 **He wrote an acclaimed book:** C. Baran-
auckas. 2001. Caryl Haskins, 93, ant expert and authority in many fields. *New
York Times,* October 13, 2001.

129 **This knowledge has been gained:** Magurran, *Evolutionary Ecology* (see Chapter Five, n. 126).

130 **In the heady days of the 1960s:** John
Endler provided many details about his
development as a scientist and the guppy
project in an email conversation March
16–June 8, 2015.

131 **This work was a great success:** J. A.
Endler. 1977. *Geographic Variation, Spe-*

ciation, and Clines. Princeton, NJ: Princeton University Press.

134 **"the field results are striking":** p. 77 in
J. A. Endler. 1980. Natural selection on
color patterns in *Poecilia reticulata. Evolution* 34: 76–91.

139 **David Reznick:** David Reznick repeatedly answered questions about his development as a scientist and the guppy
project in many emails, March 21, 2015–
November 16, 2016.

148 **but the size of the red and black spots:**
Note that in this paper (D. J. Kemp et al.,
referred to in the first endnote of this
chapter), these spots were referred to as
"orange," but they correspond to the
ones that Endler termed "red" in his
original study, so I refer to them as red
here. This study also lumped blue and
iridescent spots together, in contrast to
Endler's study.

149 **At this point, all we can say:** A different
research group compared color evolu-
tion in another of Reznick's introductions and failed to find any evidence for
color change. This study, however, did
not use the state-of-the-art color analysis brought to bear by the animal vision
experts in the Reznick and Endler study
conducted in 2005. Moreover, in a subsequent study, the other group also
failed to detect the differences that
Reznick's and Endler's team did find,
suggesting that the other group's
method wasn't capable of detecting at
least some types of color evolution (specifically, increased iridescence). The authors of the other group provide a
somewhat defensive justification of their
methods in their paper, presumably re-
butting criticisms that came up in the
review process. At this point, it's hard to
know what to make of their result.
Clearly, what is needed is further examination of other guppy introductions
using the best available methods.

150 **To find out, she collected guppies:**
S. O'Steen, A. J. Cullum, and A. F. Ben-
nett. 2002. Rapid evolution of escape

ability in Trinidadian guppies (*Poecilia reticulata*). *Evolution* 56: 776–784.

chapter six: LIZARD CASTAWAYS

155 **If you visit the Bahamas:** My book summarizes experimental work on Bahamian anoles through 2009: J. B. Losos. 2009. *Lizards in an Evolutionary Tree.* (see Ch. Two, n. 58). The key papers are: T. W. Schoener and A. Schoener. 1983. The time to extinction of a colonizing propagule of lizards increases with island area. *Nature* 302: 332–334; J. B. Losos, T. W. Schoener, and D. A. Spiller. 2004. Predator-induced behaviour shifts and natural selection in field-experimental lizard populations. *Nature* 432: 505–508; J. B. Losos et al. 2006. Rapid temporal reversal in predator-driven natural selection. *Science* 314: 1111; J. J. Kolbe et al. 2012. Founder effects persist despite adaptive differentiation: a field experiment with lizards. *Science* 335: 1086–1089.

chapter seven: FROM MANURE TO MODERN SCIENCE

181 **More than 170 years:** Determining what actually is the longest-running experiment in the world is not easy. Many online sources refer to an experiment watching tar drop through a funnel, but that study dates back only to the 1920s. Through online searches, I was unable to find any ongoing experiment that predates the Rothamsted studies. There are reports of a battery-powered bell that has been operating since 1840, just before the Rothamsted experiments began, but it's not clear to me that observing this bell constitutes an experiment.

181 **In turn, this led to experimentation:** For an overview of Rothamsted, the Park Grass Experiment, and Roy Snaydon's experiments, these references are a

good starting point: J. Silvertown. 2005. *Demons in Eden: The Paradox of Plant Diversity.* Chicago: University of Chicago Press; J. Silvertown et al. 2006. The Park Grass Experiment 1856–2006: its contribution to ecology. *Journal of Ecology* 94: 801–814; J. Storkey et al. 2016. The unique contribution of Rothamsted to ecological research at large temporal scales. *Advances in Ecological Research* 55: 3–42; R. W. Snaydon. 1970. Rapid population differentiation in a mosaic environment. I. The response of *Anthoxanthum odoratum* populations to soils. *Evolution* 24: 257–269; R. W. Snaydon and M. S. Davies. 1972. Rapid population differentiation in a mosaic environment. II. Morphological variation in *Anthoxanthum odoratum. Evolution* 26: 390–405; S. Y. Strauss et al. 2007. Evolution in ecological field experiments: implications for effect size. *Ecology Letters* 11: 199–207. My description of this work draws on conversations with Roy Snaydon (June 4–27, 2015), Stuart Davies (May 27, 2015), Jonathan Silvertown (May 19–29, 2015), and Jonathan Storkey (June 2, 2015).

184 **"the experimental ground":** p. 43 in J. B. Lawes and J. H. Gilbert. 1859. *Report of Experiments with Different Manures on Permanent Meadow Land.* London: Clowes and Sons.

184 **Let's take a walk:** The description of the plots, especially number 3, is based on J. Silvertown. *Demons in Eden* (see Chapter Seven, note 181). Additional information came from personal communication with Jonathan Storkey (June 2, 2015) and from M. J. Crawley et al. 2005. Determinants of species richness in the Park Grass Experiment. *American Naturalist* 165: 179–192.

192 **The most notable of these studies:** The following papers provide an entrée to the Silwood Park rabbit experiments: M. J. Crawley. 1990. Rabbit grazing, plant competition and seedling recruitment in acid grassland. *Journal of Applied Ecology* 27: 803–820; J. Olofsson, C. De Ma-

zancourt, and M. J. Crawley. 2007. Contrasting effects of rabbit exclusion on nutrient availability and primary production in grasslands at different time scales. *Oecologia* 150 : 582–589; N. E. Turley et al. 2013. Contemporary evolution of plant growth rate following experimental removal of herbivores. *American Naturalist* 181: S21–S34; T. J. Didiano et al. 2014. Experimental test of plant defence evolution in four species using long-term rabbit exclosures. *Journal of Ecology* 102: 584–594. Many of the details of these experiments were explained to me in conversations with Marc Johnson (May 29–December 10, 2015), Mick Crawley (May 29–30, 2015), and Nash Turley (May 17–29, 2015).

194 **For example, Marc Johnson:** A. A. Agrawal et al. 2012. Insect herbivores drive real-time ecological and evolutionary change in plant populations. *Science* 338: 113–116.

194 **Other studies have looked:** T. Bataillon et al. 2016. A replicated climate change field experiment reveals rapid evolutionary response in an ecologically important soil invertebrate. *Global Change Biology* 22: 2370–2379; V. Soria-Carrascal et al. 2014. Stick insect genomes reveal natural selection's role in parallel speciation. *Science* 344: 738–742.

chapter eight: EVOLUTION IN SWIMMING POOLS AND SANDBOXES

195 **These are his pools:** My description of the development of the stickleback experimental evolution research program is based primarily on conversations with Dolph Schluter (June 12, 2015–September 23, 2016), but also with Rowan Barrett (March 4, 2015–July 12, 2016) and Diana Rennison (September 23–November 11, 2016).

196 **His detailed studies are now classics:** D.

Schluter, T. D. Price, and P. R. Grant. 1985. Ecological character displacement in Darwin's finches. *Science* 227: 1056–1059.

200 **The answer was yes:** R. D. H. Barrett, S. M. Rogers, and D. Schluter. 2008. Natural selection on a major armor gene in threespine stickleback. *Science* 322: 255–257; R. D. H. Barrett et al. 2011. Rapid evolution of cold tolerance in stickleback. *Proceedings of the Royal Society of London B* 278: 233–238.

201 **Working with leading genome experts:** P. F. Colosimo et al. 2005. Widespread parallel evolution in sticklebacks by repeated fixation of *Ectodysplasin* alleles. *Science* 307: 1928–1933.

202 **selection for pelvic spine length:** D. J. Rennison 2016. Detecting the drivers of divergence: identifying and estimating natural selection in threespine stickleback. Ph.D. dissertation, University of British Columbia.

205 **Bird predation:** L. R. Dice. 1947. Effectiveness of selection by owls of deermice (*Peromyscus maniculatus*) which contrast in color with their background. *Contributions from the Laboratory of Vertebrate Biology* 34: 1–20.

205 **inference drawn from genetic differences:** C. R. Linnen et al. 2009. On the origin and spread of an adaptive allele in deer mice. *Science* 325: 1095–1098.

205 **Nebraska Sandhills:** My description of the Sandhills deer mouse project is based on extensive conversations with Rowan Barrett, June 5, 2015–July 12, 2016.

chapter nine: REPLAYING THE TAPE

229 **On February 24, 1988:** Lenski's experiments are well chronicled and it is easy to find popular articles on them online or in magazines, such as T. Appenzeller. 1999. Test tube evolution catches time in a bottle. *Science* 284: 2108; E. Pennissi. 2013. The man who bottled evolution.

Science 342: 790–793. Lenski has summarized the results of the work himself on his blog *Telliamed Revisited* (December 29, 2013, https://telliamedrevisited. wordpress.com/2013/12/29/ what-weve-learned-about-evolution-from-the-ltee-number-5/6). I learned of many of the details of this research program and the personal histories of many involved thanks to a visit to the Lenski Lab (October 2–3, 2014) and to extensive correspondence with Zack Blount (December 20, 2014–November 6, 2016), as well as much communication with Rich Lenski (August 17–27, 2015), Tim Cooper (January 24–27, 2015), and Chris Borland (February 18–23, 2015).

233 **Proceedings of the National Academy of Sciences:** R. E. Lenski and M. Travisano. 1994. Dynamics of adaptation and diversification: a 10,000-generation experiment with bacterial populations. *Proceedings of the National Academy of Sciences of the United States of America* 91: 6808–6814.

233 **Differences still existed:** p. 240 in R. E. Lenski. 2004. Phenotypic and genomic evolution during a 20,000-generation experiment with the bacterium *Escherichia coli*. *Plant Breeding Reviews* 24, pt. 2: 225–265.

234 **"evolution was pretty repeatable":** p. 32 in R. E. Lenski. 2011. Evolution in action: a 50,000-generation salute to Charles Darwin. *Microbe* 6: 30–33.

234 **Like Lenski, Paul Rainey:** Email conversations with Paul Rainey (February 15, 2015–March 17, 2015) provided the details on his background and that of the *Pseudomonas fluorescens* research program.

239 **Travisano and Rainey finished up:** P. B. Rainey and M. Travisano. 1998. Adaptive radiation in a heterogeneous environment. *Nature* 394: 69–72.

244 **However, in the snowflakes:** W. C. Ratcliff et al. 2012. Experimental evolution of multicellularity. *Proceedings of the National Academy of Sciences of the United States of America* 109: 1595–1600.

245 **Another study on *E. coli*:** O. Tenaillon et al. 2012. The molecular diversity of adaptive convergence. *Science* 335: 457–461.

chapter ten: BREAKTHROUGH IN A BOTTLE

251 **"named after places in myth":** Zack Blount, personal communication, March 13, 2015.

252 **"the most intensely studied species":** C. Zimmer. 2012. The birth of the new, the rewiring of the old. *The Loom*, September 19, 2012, http://blogs.discovermagazine.com/ loom/2012/09/19/the-birth-of-the-new-the-rewiring-of-the-old/#.WCO6JeErJjs.

253 **"The bad news is":** p. 48 in Gould, *Wonderful Life* (see Introduction; n. 17).

254 **"earnest and smart":** Rich Lenski, personal communication, August 17, 2015.

256 **Two results are clear:** Z. D. Blount, C. Z. Borland, and R. E. Lenski. 2008. Historical contingency and the evolution of a key innovation in an experimental population of *Escherichia coli*. *Proceedings of the National Academy of Sciences of the United States of America* 105: 7899–7906.

257 **working his molecular magic:** Z. D. Blount et al. 2012. Genomic analysis of a key innovation in an experimental *Escherichia coli* population. *Nature* 489: 513–518.

258 **there is a second mutation:** E. M. Quandt et al. 2015. Fine-tuning citrate synthase flux potentiates and refines metabolic innovation in the Lenski evolution experiment. *eLife* 4: e09696.

259 **"The tension between chance":** J. Dennehy. 2008. This week's citation classic: the fluctuation test. *The Evilutionary Biologist*, July 9, 2008, http://evilutionarybiologist.blogspot.com/2008/07/this-weeks-citation-classic-fluctuation.html.

260 **Recent work in Paul Rainey's lab:** P. A. Lind, A. D. Farr, and P. B. Rainey. 2015. Experimental evolution reveals hidden diversity in evolutionary pathways. *eLife* 4: e07074.

261 **another *E. coli* study:** M. L. Friesen et

al. 2004. Experimental evidence for sympatric ecological diversification due to frequency-dependent competition in *Escherichia coli. Evolution* 58: 245–260.

262 **In an interview:** D. S. Wilson. 2016. Evolutionary biology's master craftsman: an interview with Richard Lenski. *This View of Life,* May 30, 2016, https://evolution-institute.org/article/evolutionary-biologys-master-craftsman-an-interview-with-richard-lenski/.

chapter eleven: JOTS, TITTLES, AND DRUNKEN FRUIT FLIES

263 **"I call this experiment":** p. 48 in Gould, *Wonderful Life* (see Introduction; n. 17).

264 **"This magnificent ten-minute scene":** Ibid., p. 287.

264 **". . . any replay":** Ibid., p. 289.

267 **"Historical explanations take the form":** Ibid., p. 283.

269 **"Alter any early event":** Ibid., p. 51.

270 **"go back in time":** J. Maynard Smith. 1992. Taking a chance on evolution. *New York Review of Books,* May 14, 1992.

272 **More to the point:** Fred Cohan kindly detailed the backstory for this project in a series of emails February 19, 2015–November 6, 2016.

275 **The average fly:** F. M. Cohan and A. A. Hoffman. 1986. Genetic divergence under uniform selection. II. Different responses to selection for knockdown resistance to ethanol among *Drosophila melanogaster* populations and their replicate lines. *Genetics* 114: 145–163.

276 **A similar experiment:** A. Spor et al. 2014. Phenotypic and genotypic convergences are influenced by historical contingency and environment in yeast. *Evolution* 68: 772–790.

278 **Reporting this result:** R. E. Lenski et al. 1991. Long-term experimental evolution in *Escherichia coli.* I. Adaptation and divergence during 2,000 generations. *American Naturalist* 138: 1315–1341.

279 **To test this idea:** M. Travisano et al. 1995. Experimental tests of the roles of adaptation, chance, and history in evolution. *Science* 267: 87–90.

282 **Gould really messed things up:** J. Beatty. 2006. Replaying life's tape. *Journal of Philosophy* 103: 336–362.

282 **Surprisingly few studies:** Surprisingly, so far no one has written a comprehensive review of such studies. The closest to date are V. Orgogozo. 2015. Replaying the tape of life in the twenty-first century. *Interface Focus* 5: 20150057, and a book chapter written by Zack Blount that focuses on experiments with microbes and provides a nice overview of the LTEE: Z. B. Blount. 2016. History's windings in a flask: microbial experiments into evolutionary contingency. Pp. 244–263 in G. Ramsey and C. H. Pence, eds., *Chance in Evolution.* Chicago: University of Chicago Press.

chapter twelve: THE HUMAN ENVIRONMENT

286 **such an experiment has been conducted:** A. Wong, N. Rodrigue, and R. Kassen. 2012. Genomics of adaptation during experimental evolution of the opportunistic pathogen *Pseudomonas aeruginosa. PLoS Genetics* 8: e1002928.

288 **The Danish researchers:** R. L. Marvig et al. 2014. Convergent evolution and adaptation of *Pseudomonas aeruginosa* within patients with cystic fibrosis. *Nature Genetics* 47: 57–64.

290 **The biochemical function:** Some of the precise details here and earlier were provided by Rasmus Marvig, personal communication, July 17, 2015 and May 22, 2016.

291 **Another study yielded:** T. D. Lieberman et al. 2011. Parallel bacterial evolution within multiple patients identifies candidate pathogenicity genes. *Nature Ge-*

netics 43: 1275–1280.

292 **An international team:** M. R. Farhat et al. 2013. Genomic analysis identifies targets of convergent positive selection in drug-resistant *Mycobacterium tuberculosis. Nature Genetics* 45: 1183–1189.

294 **"even a modest predictive power":** p. 243 in A. C. Palmer and R. Kishony. 2013. Understanding, predicting and manipulating the genotypic evolution of antibiotic resistance. *Nature Reviews Genetics* 14: 243–248.

295 **In this experiment, our old friend:** E. Toprak et al. 2012. Evolutionary paths to antibiotic resistance under dynamically sustained drug selection. *Nature Genetics* 44: 101–106.

297 **phenotypic differences and non-convergence:** Y. Stuart et al. Contrasting effects of environment and genetics generate a predictable continuum of parallel evolution. *Nature Ecology & Evolution,* accepted for publication.

297 **Some scientists are even suspicious:** A. E. Lobkovsky and E. V. Koonin. 2012. Replaying the tape of life: quantification of the predictability of evolution. *Frontiers in Genetics* 3(246): 1–8.

298 **"is a weak form":** p. 484 in J. A. G. M. de Visser and J. Krug. 2014. Empirical fitness landscapes and the predictability of evolution. *Nature Reviews Genetics* 15: 480–490.

299 **the diversity of mutations:** H. Jacquier et al. 2013. Capturing the mutational landscape of the beta-lactamase TEM-1. *Proceedings of the National Academy of Sciences of the United States of America* 110: 13067–13072.

301 **"tape of life":** p. 113 in D. M. Weinreich et al. 2006. Darwinian evolution can follow only very few mutational paths to fitter proteins. *Science* 312: 111–114.

301 **In the free market of mutations:** M. L. M. Salverda et al. 2011. Initial mutations direct alternative pathways of protein evolution. *PLoS Genetics* 7: e1001321.

305 **a number of mosquito species:** N. Liu. 2015. Insecticide resistance in mosqui-

toes: impact, mechanisms, and research directions. *Annual Review of Entomology* 60: 537–559.

305 **more than thirty different insect species:** R. H. ffrench-Constant. 2013. The molecular genetics of insecticide resistance. *Genetics* 194: 807–815; F. D. Rinkevich, Y. Du, and K. Dong. 2014. Diversity and convergence of sodium channel mutations involved in resistance to pyrethroids. *Pesticide Biochemistry and Physiology* 106: 93–100.

305 **The amount of farmland:** B. E. Tabashnik, T. Brévault, and Y. Carrière. 2013. Insect resistance to Bt crops: lessons from the first billion acres. *Nature Biotechnology* 31: 510–521. In addition, Bruce Tabashnik provided some updated figures (personal communication, October 13, 2016).

305 **resistance to one type of Bt toxin:** Y. Wu. 2014. Detection and mechanisms of resistance evolved in insects to Cry toxins from *Bacillus thuringiensis. Advances in Insect Physiology* 47: 297–342.

305 **seven caterpillar species:** B. Tabashnik. 2015. ABCs of insect resistance to Bt. *PLoS Genetics* 11: e1005646.

306 **One particularly well-understood case:** N. M. Reid et al. 2016. The genomic landscape of rapid repeated evolutionary adaptation to toxic pollution in wild fish. *Science* 354: 1305-1308.

307 **Humans also impose:** A good place to start on this topic are two reviews: F. W. Allendorf and J. J. Hard. 2009. Human-induced evolution caused by unnatural selection through harvest of wild animals. *Proceedings of the National Academy of Sciences of the United States of America* 106: 9987–9994; M. Heino, B. Díaz Pauli, and U. Dieckmann. 2015. Fisheries-induced evolution. *Annual Review of Ecology, Evolution and Systematics* 46: 461–480.

308 **The largest Atlantic cod:** Allendorf and Hard. Human-induced evolution (see Chapter Twelve, note 307), 9987–9994; Doug Swain (personal communication, October 11 and October 25, 2016) and Loretta O'Brien (personal commu-

nication, October 24, 2016).

conclusion: FATE, CHANCE, AND THE INEVITABILITY OF HUMANS

311 **The luxuriant vegetation:** See M. Wilhelm and D. Mahison. 2009. *James Cameron's Avatar: An Activist Survival Guide.* New York: HarperCollins; Pandorapedia: the official field guide. https://www.pandorapedia.com/pandora_url/dictionary.html; *Pandora Discovered*, a four-minute film narrated by Sigourney Weaver, is very instructive as well, https://www.youtube.com/watch?v=GBGDmin_38E#t=93.

312 **"what we see here [on Earth]":** p. 566 in S. Conway Morris. 2011. Predicting what extra-terrestrials will be like: and preparing for the worst. *Philosophical Transactions of the Royal Society A* 369: 555–571.

316 **"Some people—science fiction writers":** p. 29 in C. Sagan. 1980. *Cosmos.* New York: Random House.

318 **More liberal estimates:** S. B. Carroll. 2001. Chance and necessity: the evolution of morphological complexity and diversity. *Nature* 409: 1102–1109; K. J. Niklas. 2014. The evolutionary-developmental origins of multicellularity. *American Journal of Botany* 101: 6–25.

318 **Elephants placing boxes:** F. de Waal. 2016. *Are We Smart Enough to Know How Smart Animals Are?* New York: W. W. Norton.

320 **a population of chimpanzees:** M. Roach. 2008. Almost human. *National Geographic* 213(4): 124–144.

322 **"Much too human":** Thomas Holtz quoted in J. Hecht. 2007. Smartasaurus. *Cosmos.* July 9, 2007, https://cosmosmagazine.com/palaeontology/smartasaurus.

322 **"suspiciously human":** D. Naish. 2012. Dinosauroids revisited, revisited. *Tetrapod Zoology,* October 27, 2012, https://blogs.scientificamerican.com/tetrapod-zoology/dinosauroids-revisited-revisited/.

322 **Conway Morris later agreed:** quoted in G. Hatt-Cook. 2007. What if the asteroid had missed? *BBC News,* March 13, 2007, http://news.bbc.co.uk/2/hi/science/nature/6444811.stm.

322 **"We invite our colleagues":** p. 36 in Russell and Séguin. Reconstructions of the small Cretaceous theropod *Stenonychosaurus inequalis* (see Introduction, note 6).

323 **"evolved into a humanoid":** Thinkaboutit's Alien Type Summary—Dinosauroids. *Thinkaboutit.* Accessed June 1, 2016. http://www.thinkaboutit-aliens.com/think-aboutits-alien-type-summary-dinosauroids/. In the quote, I changed "amphibian" to "reptile" for the sake of accuracy.

323 **Amidst all this nonsense:** Naish. Dinosauroids revisited, revisited (see Conclusion, n. 322).

325 **Adorned with feathers:** S. Roy. 2016. *Deviant Art.* Accessed November 12, 2016. http://povorot.deviantart.com/gallery/9348116/The-Dinosauroids.

327 **Harry Burrell:** H. Burrell. 1927. *The Platypus.* Sydney: Angus & Robertson.

327 **tactile and electric stimuli:** J. D. Pettigrew. 1999. Electroreception in monotremes. *Journal of Experimental Biology* 202: 1447–1454; T. Grant. 2008. *Platypus,* 4th ed. Collingwood, Australia: CSIRO Publishing.

327 **The platypus, like humans:** Jerry Coyne insightfully discusses evolutionary singletons (and uses both that term and "one-offs"). He also makes a point about the elephant similar to the one here about the platypus. See J. A. Coyne. 2015. Simon Conway Morris's new book on evolutionary convergence. Does it give evidence for God? *Why Evolution Is True,* February 8, 2015, https://whyevolutionistrue.wordpress.com/2015/02/08/simon-conway-morriss-new-book-on-evolutionary-convergence-does-it-give-evidence-for-god/; J. A. Coyne. 2016. *Faith versus Fact: Why Science and Religion Are Incompatible.* New York: Penguin.

331 **Edward O. Wilson:** pp. 113–117 in E. O. Wilson. 2015. *The Meaning of Human Existence.* New York: Liveright.

Index

Italicized Numbers Refer to Pages with Illustrations

Printed in the United States
by Baker & Taylor Publisher Services